An Introduction to Satellite Television

(a guide to the technicalities of space-age television)

REVISED EDITION

ALSO OF INTEREST

A TV-DXERS HANDBOOK **BP176**

R. Bunney

The author discusses the possibilities and problems of receiving television signals over long distances and the resolving of such pictures with the minimum of distortion on the TV screen. This, in essence, is the fascinating hobby of TV-DXing. Roger Bunney is one of the leading authorities in the country on this subject and an active TV-DXing enthusiast.

He has now extensively revised, enlarged and completely updated his previous works to produce this new book.

The satellite TV section has been greatly expanded to encompass further generalised information, since in the coming decade 'direct-to-home' transmissions will perhaps herald the most dramatic changes ever in the broadcasting field.

Included are many units and devices which have been designed by experienced enthusiasts and often considerable ingenuity and thought have gone into their practical development to overcome individual problems.

It is hoped that the accumulated information in this book will be a practical guide for the beginner and a source of reference for the established enthusiast.

0 85934 150 X *96 pages* *264 x 195 mm* *1986* **£5.95**

An Introduction To Satellite Television

by
F. A. Wilson
CGIA, CEng, FIEE, FBIM.

BERNARD BABANI (publishing) LTD
THE GRAMPIANS
SHEPHERDS BUSH ROAD
LONDON W6 7NF
ENGLAND

PLEASE NOTE

Although every care has been taken with the production of this book to ensure that any projects, designs, modifications and/or programs etc. contained herewith, operate in a correct and safe manner and also that any components specified are normally available in Great Britain, the Publishers do not accept responsibility in any way for the failure, including fault in design, of any project, design, modification or program to work correctly or to cause damage to any other equipment that it may be connected to or used in conjunction with, or in respect of any other damage or injury that may be so caused, nor do the Publishers accept responsibility in any way for the failure to obtain specified components.

Notice is also given that if equipment that is still under warranty is modified in any way or used or connected with home-built equipment then that warranty may be void.

© 1988 and © 1989 BERNARD BABANI (publishing) LTD

First Published – March 1988
Revised Edition – January 1989

British Library Cataloguing in Publication Data:
Wilson, F. A.
An introduction to satellite television.
1. Direct broadcast satellite television —
Equipment and supplies
I. Title
621.388'5 TK6677

ISBN 0 85934 169 0

Printed and Bound by The Guernsey Press Co. Ltd, Channel Islands

PREFACE

Through pathless realms of Space
Roll on!

Sir W. S. Gilbert

In the beginning for many of us the only experience of satellite transmission was the occasional glimpse on tv of an enormous "dish" antenna (or aerial) pointing skywards. But time and technology go hand in hand and now we ourselves can have our own private dishes, collect a tiny share of the signals from a far-off satellite and hopefully enjoy the tv programmes on offer. Moreover we can peek into other countries' tv affairs. The technique is fascinating indeed but let us be under no illusions, it is complicated and one which is undergoing unprecedented development. Yet without burdening ourselves too much with the electronic technology involved, it is possible to gain an all-round awareness of the whole system. After all, in a restaurant it is hardly necessary to understand the chemistry of cooking to be able to enjoy the food. We can then undertake installation of our own home equipment or certainly know what to expect from a dealer who does it for us. Before even considering installation however comes the overriding question as to whether it is possible to join the satellite fraternity at all. Happily the number of homes excluded is rapidly decreasing now that dishes are small enough for chimney-stack mounting.

The full story from rocket launch to tv screen involves a multitude of mechanical and electronic disciplines as a glance at the Table of Contents shows. The discussion therefore covers a wide range. Striking a balance between including just sufficient for the *modus operandi* to be understood and on the other hand saturating the reader with detail is not easy. Let us hope that we have got it right. However, the many engineers and scientists who require more than just an insight into the whole affair are not forgotten. They delight in what mathematical formulae have to tell so for these dedicated people the related formulae and computer programs are included in the Appendices. In this way the book attempts to cater for all tastes and all that is required of the reader is that the grey matter is in good working order.

It must be emphasized that this is a book on *how* it works, not on tv programmes and programme contractors, these are the province of newspapers and magazines.

Satellite communication is new, much of it very new. Accordingly to be up to date we avoid the partially defunct Imperial System (pints, pounds, inches, miles) and hitch our wagon to the metric.

Some updating may be appropriate on the struggle for supremacy between the American *antenna* and the English *aerial*. The English are losing the battle so there are now fewer aerials and more antennas. The word is derived from Latin and one would expect the plural to be antennae but no, these only refer to our creepy-crawly friends. To add to the confusion satellite tv has its *dishes*. A dish is an antenna in a more colloquial form so antenna or dish is used as seems appropriate. The two words may seem to be synonymous but although a dish is always an antenna, an antenna is not necessarily a dish.

Raised references will be found in the main text, for example, " . . . seconds.[A2(2)]" Here the A2 indicates Appendix 2 while the bracketed 2 refers to equation 2 of that Appendix. Alternatively A2.2 refers to the second section in Appendix 2. Where an Appendix is a technical extension of a Chapter, the Appendix has the same number as the Chapter.

As in all professions, some jargon is inevitable so particularly useful is Appendix 10 which contains a glossary of terms.

F. A. Wilson, CGIA, CEng, FIEE, FBIM

CONTENTS

Chapter 1

THE SATELLITE SYSTEM

Never let us think that the idea of communication satellites is new. Way back in 1945 the famous American science writer, Arthur C. Clarke, actually suggested that "extra-terrestrial relays" were a possibility. The prognostications of such men of vision so often come true and once again technology has advanced sufficiently for the original ideas to bear fruit. All thanks to the space programmes of those countries anxious to keep abreast of the times.

As we go we will dispel much of the mumbo-jumbo of communications engineering. Right now let us be sure of what a *satellite* actually is. The name originally referred to a heavenly body revolving round a planet, for example the Moon is a satellite of the Earth and has been running around its lord and master since time immemorial. In its turn the Earth is a satellite of the Sun. But now we have *artificial* satellites, boxes of complicated equipment continually flying round the Earth and even round other planets. The description "artificial" is now usually omitted.

1.1 The Overall Picture

Before plunging into the mysteries of satellites, let us first see the overall picture, that is, get some idea of what they are, where they are and what they do. Once conversant with the system as a whole, it then becomes easier to look at the many parts to see how they all fit together.

The earlier *communications* satellites (for telephony, tv and data) travelled high up around the Earth and the earthly antennas had to follow or "track" them. Pointing a large, steerable antenna to an invisible, swiftly moving object in the sky is not the easiest of tasks and would certainly not be viable for home use. Fortunately as both space and electronic technologies have developed, the difficulty has now been side-stepped.

Satellites for relaying tv programmes direct into people's homes are now in *geostationary* orbits. *Geo* comes from the Greek, meaning "earth" so leading to the concept of a satellite stationary *with respect to* Earth. Accordingly, for such a satellite the antennas remain in fixed positions and no tracking is required.

The question immediately arises – if the Earth is moving around the Sun and also spinning on its axis, how can anything in space become geostationary, i.e. always visible from the same earthly point? Figure 1.1 shows how simple the principle is. The satellite travels round at a very fast speed of some 11,000 kilometres per hour (km/h, nearly, 7,000 miles per hour) and at a height of nearly 36,000 km (some 23,000 miles). In fact Concorde, travelling straight up at the speed of sound would take about 30 hours

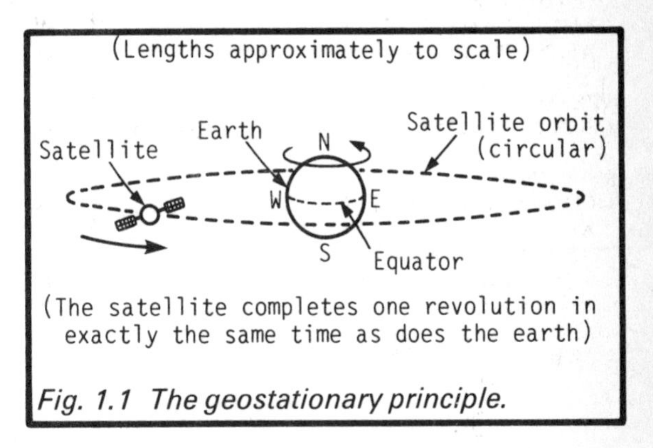

Fig. 1.1 The geostationary principle.

to reach one. At this high speed up in the heavens the satellite goes round the Earth in a circle (its *orbit*) once in about 24 hours as does the Earth itself. Accordingly, from the Earth the satellite appears to be stationary in the sky even though in fact both an Earth-bound observer and the satellite are travelling fast. It is all a question of relativity as Einstein pointed out.

The principle can be illustrated by a cycle wheel. The hub represents the Earth and the rim the satellite orbit. Any point on the hub remains opposite to some point on the rim because both hub and rim revolve together, the spokes make sure of that. In a satellite system other, not so tangible devices of Nature are employed instead of spokes but they produce the same effect. Because of Arthur C. Clarke's foresight, the geostationary orbit is frequently referred to as the *Clarke Belt.*

At this early stage we can look upon the satellite itself as being many things but mainly a tv receiver tuned to signals shot up to it from the Earth from one of the large "dishes" which adorn our landscape. The programme is boosted up in power and relayed back to Earth to be picked up by the many smaller dishes on roofs and in gardens.

A spotlight in the theatre starts off with a small, powerful source of light which is focussed into a narrow beam by a reflector. On reaching the stage an elongated pool of light is formed. Satellite transmission is similar, light and radio waves have much in common so a radio beam is focussed by the satellite in the same way by a reflector or *dish* and it arrives on Earth to "illuminate" the area aimed at. This could be for example the whole of the United Kingdom or France. The area is known as the satellite *footprint.* Direct Broadcasting by Satellite (DBS) is of this form. These satellites are specifically designed for the purpose of broadcasting tv and are powerful enough to enable reasonably small dishes to be used. As time goes on more and more of the DBS variety will join in thereby greatly increasing the

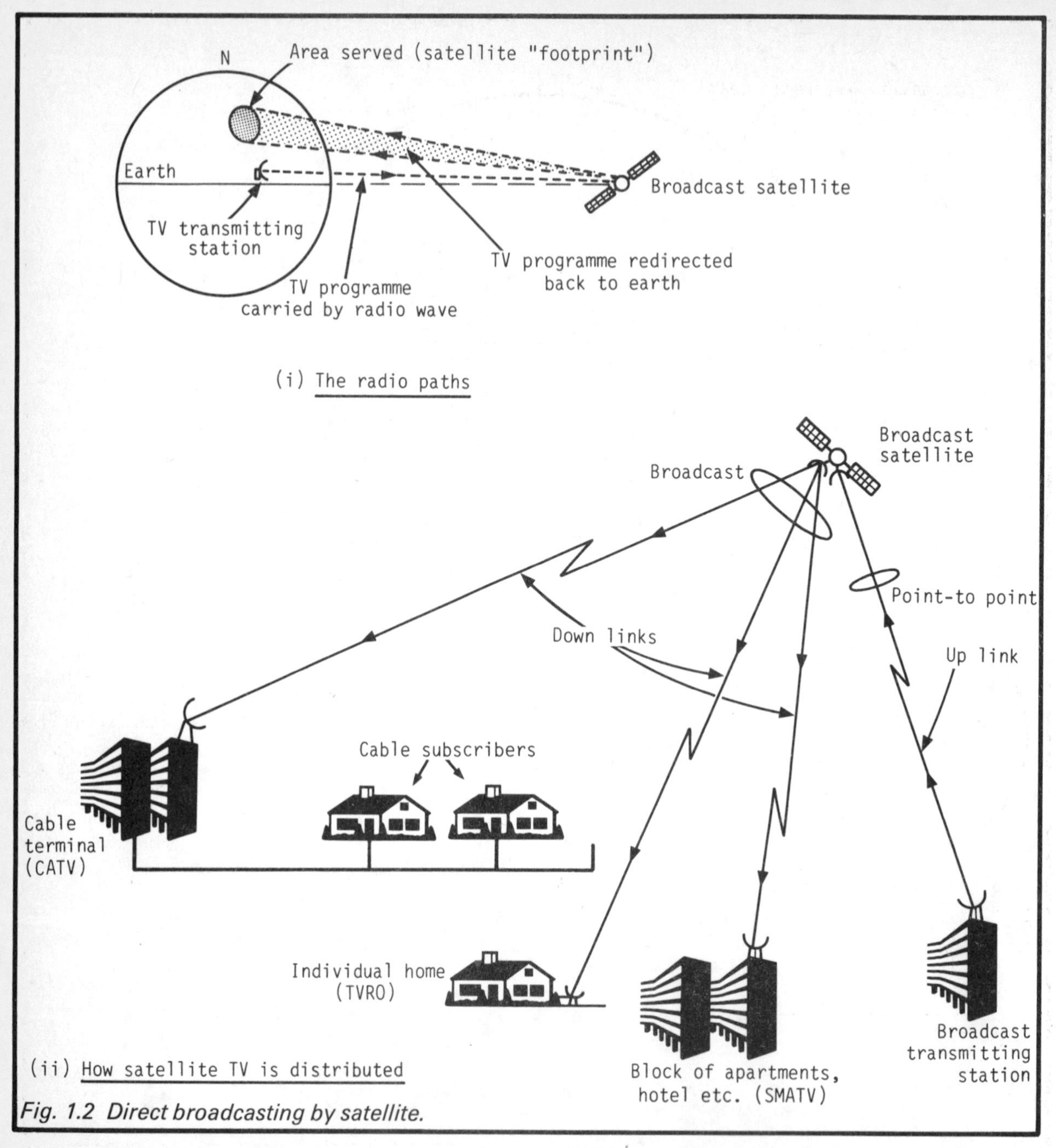

Fig. 1.2 Direct broadcasting by satellite.

number of tv channels available. Happily we are not restricted to the favours of one satellite only for the more adventurous can move their dishes to pick up other satellites, or even have a motor to do it for them.

Figure 1.2 illustrates the basic arrangement. In (i) the transmitting station is shown remote from the area served, but it can in fact be anywhere within the area. In (ii) of the Figure are shown the various types of user of satellite tv. Where a receiving dish feeds signals to a block of apartments, hotel, group of houses etc., the installation is classified as SMATV (Satellite to Master Antenna TV). Cable terminals receive the programmes and distribute them directly to their subscribers who pay for the service instead of installing their own receiving equipment. This arrangement is known as CATV (Community Antenna Television). For the individual home it is TVRO (Television Receive Only).

We may feel that we are already getting bogged down with terms and their abbreviations. Don't worry, there is always Appendix 10 for reminders.

1.1.1 The Steps Forward

The first operational satellite ever to be used was the Moon. In the late fifties the technique of reflecting radio waves from this natural satellite began to get under way. Materials can reflect radio waves just as a

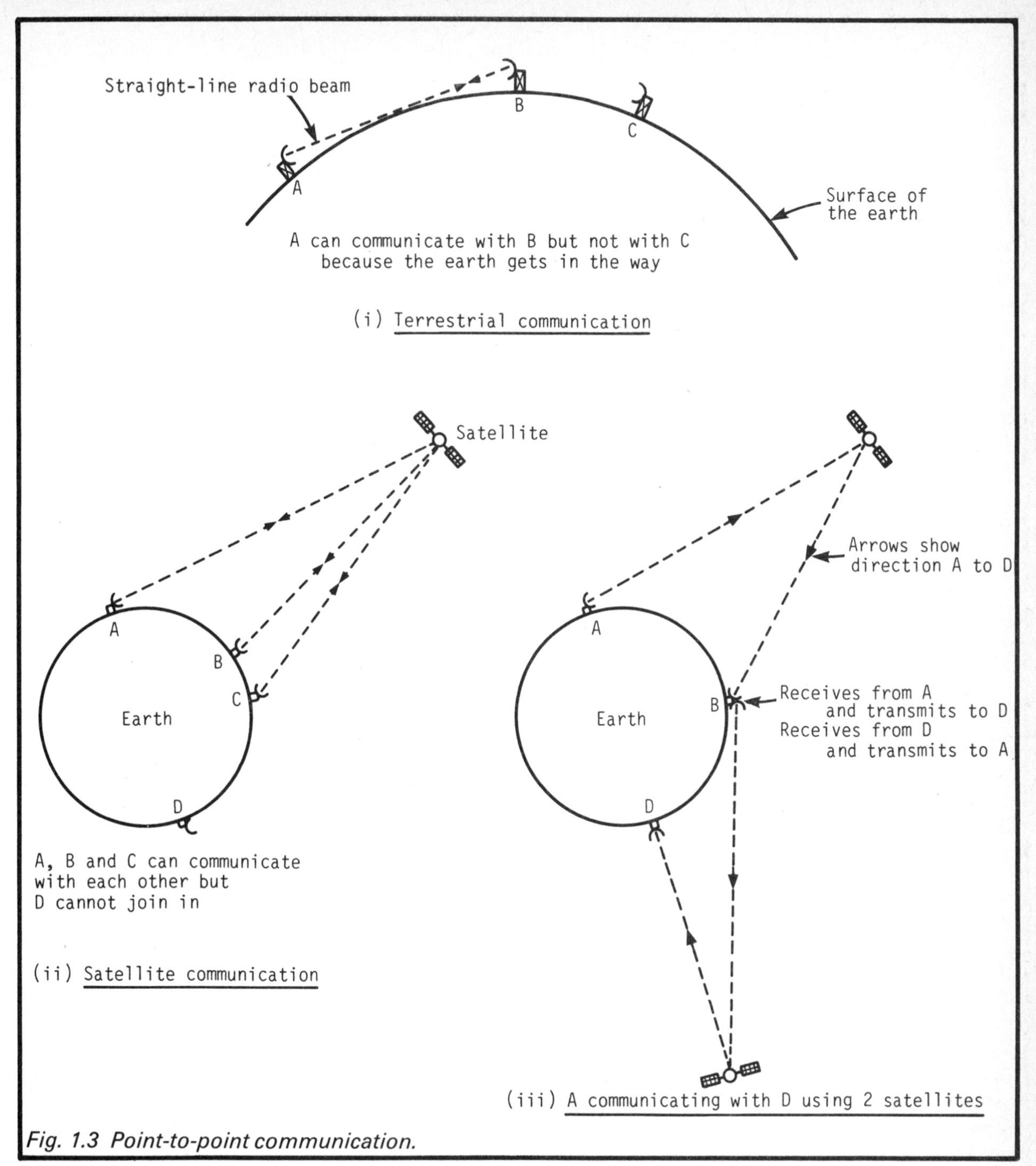

Fig. 1.3 Point-to-point communication.

mirror reflects light although the Moon probably absorbed more than it reflected back. Certainly with the Moon there were no launching problems and it is large enough not to be missed bearing in mind that it is moving relative to the Earth. The US Navy and a host of radio amateurs did their best but "moon-bounce" as it was then called, inevitably gave way to the growing band of artificial satellites.

In 1957 the Russians put up SPUTNIK I, the first. Then followed SPUTNIK II, the USA EXPLORER I, NASA's SCORE, followed by the *passive* ECHO I. Passive means that there are no *active* devices such as amplifiers to boost the signals, merely doing nothing more than reflecting them. *Active* devices need electrical power to do their job, passive do not. In fact ECHO was simply a large balloon with an aluminium coating, this reflected radio waves back to Earth and much more efficiently than did the Moon.

Soon active communication satellites came along, starting with TELSTAR I (1962) followed by a second one and then RELAY I and II. Signals sent up to these were made more powerful (boosted) and then directed back to Earth. These showed great promise but were all limited to low altitudes because rockets of the time could not carry them higher. As soon as they could, the USA SYNCOM III was placed in the first geostationary orbit (1964) and from then, satellite communication systems grew apace. INTELSAT I (Early Bird) to INTELSAT VI, SATCOM (USA), WESTAR (USA), ANIK (Canada), EUTELSAT (Europe), ARABSAT (Arab consortium) to name only a few and still they come. The orbit is slowly but surely filling up.

1.2 Why Satellites?

With the little bit of information so far, one might well ask why a satellite is needed at all. Is it not possible to shoot straight from transmitting dish to the receiving dishes instead of up into the sky and back again? Come to that, what is wrong with the tv system we already have, merely broadcast stations with everybody's antennas pointing to them? The first question would not arise if the Earth were flat but this idea lost favour many years ago. Figure 1.3 at (i) shows the limitations of *point-to-point* communication with a round Earth. At (ii) the Figure shows how the distance limitation is overcome using a satellite because A can now communicate with both B and C. Note however that D is not catered for in this arrangement. The reason is that point-to-point communication employs straight line radio beams and it is possible for the Earth to effectively block the path. But for such world-wide working the difficulty is overcome by using more than one satellite so at (iii) is seen how A reaches D using two "hops". There could of course be a *direct* link between the two satellites, thereby reducing the total journey but this is not without its own special problems.

We can understand the limitations of Earth-bound tv systems by considering the radio waves they employ. These particular ones have an undesirable habit of getting absorbed by the ground and other solid objects as they travel away from the transmitting station. After no more than a few tens of kilometres the signal becomes too weak to do its job, therefore many transmitting stations are needed, each covering a small area. In fact for the UK over 600 such stations are required for reasonably complete coverage of the country. On the other hand, a single transmitting station with one satellite can do the same job. What is more, at home we can realign our dishes to pick up different satellites and as time goes on there will be many of these beaming tv programmes down. In addition satellites are able to serve those small areas where terrestrial broadcasting has difficulty. Finally some desirable improvements in picture quality are now possible and these can be introduced as the satellite system expands.

Many home stations are already working via the existing satellites, e.g. INTELSAT and EUTELSAT. Compared with the more modern newcomers these are low powered and need relatively large dishes. Such antennas are viable for cable distribution networks and also for home use if a 1.2 to 1.8 metre diameter dish can be tolerated. As more powerful DBS satellites become available so dish size decreases and it is envisaged that most for home use will be more manageable at some 30 centimetres (about 12 inches) up to one metre diameter.

There is no doubt that the advent of satellites considerably expands the programme horizons and the great attraction for European operators is that there are over 120 million homes in the area, each a *potential* user. The uncertainty is just how many will become users and when. Behind all this lie the language and cultural barriers and the enormous cost of providing the service. But there is little doubt that slowly DBS will take its place alongside the present system and dishes and fishbone type antennas will sit side by side. It might even be that satellites will take over completely but surely such a prospect takes us well into the 21st century.

1.3 The System in Outline

An old peep-show rhyme says it all for us, "You pays your money and you takes your choice" – but it also depends on how much you pays. The least expensive receiving installation is based on a fixed dish lined up on one satellite only. For a little more the dish can be adjustable by hand to receive other satellites. At the top of the financial scale is the remote-controlled system with which a motorized dish can be made to change its outlook according to signals sent to it by twiddling knobs and switches indoors. On the very latest systems even knob twiddling is avoided, the indoor equipment can be pre-programmed to turn the dish and select the channel.

Somewhere outside then there is the dish with its clear, unobstructed view of the satellite(s). It collects the radio wave sent down and directs it into a funnel-shaped device usually fixed in front on a tripod. The funnel feeds the wave into an LNB (or LNC, but usually the former), a *low noise block converter* which changes the wave so that it is suitable for transmitting onwards through a special cable into the premises (see Figure 1.4). Choosing the most suitable dish size is a bit of a teaser. Many would-be viewers, knowing so little about the complexities of the system and possibly without even fully understanding what they want, will go straight to a dealer and put themselves in his or her hands. Not for us though for later in the book (Chapter 8) we concentrate on this very problem.

The cable from the LNB is similar to the standard circular cable which at present finds its way from the tv antenna on the roof, or in the attic, into the back of the set. This cable is not connected directly to the normal tv set but instead to a *satellite receiver* which enables the user to select the programme. The output from the receiver is in a suitable form for connection directly to the domestic tv set which can be switched

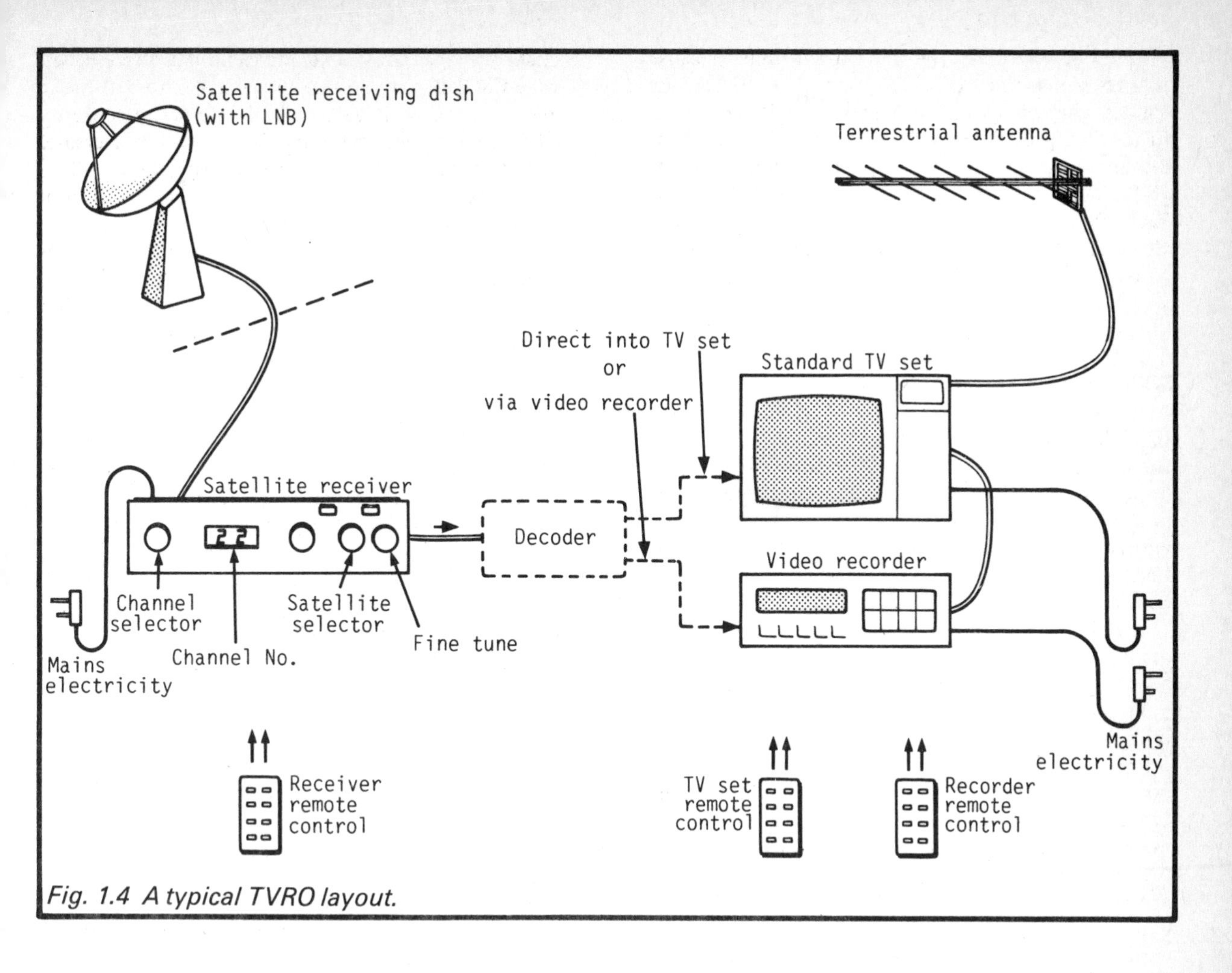

Fig. 1.4 A typical TVRO layout.

to "terrestrial" tv stations or satellite ones as required by its normal remote control or channel selection buttons. As mentioned above, other features in the receiver may include control of the dish outside to aim it at different satellites.

Figure 1.4 includes a video recorder which is unaffected by the addition of satellite channels. Note the proliferation of remote-control units, already it is possible to have one single unit take over.

1.3.1 Paying Up

We cannot expect to receive all the extra tv channels free. People who distribute the television programmes (the programme contractors) have to pay those who actually make them so they need to get their money back and with luck, some profit. Taxpayers' money may not be available nor is it envisaged that advertising will always generate sufficient income, hence payment has to be made for certain channels in some other way. Accordingly the technique of *encryption* is likely to be used.

We may recall the *scrambled* telephone in which speech is mutilated before being sent onwards. On its journey therefore anybody listening in is unable to understand the message. At the distant end the speech is descrambled so that the legitimate listener hears the message clearly. Early scramblers made disapproved listening-in difficult, modern ones make it almost impossible. *Encryption* is the same technique used with tv, in fact encryption and scrambling are virtually synonymous. On an encrypted channel the signal is corrupted in such a way that it no longer produces a worth-while picture unless a descrambler (or decoder, see Figure 1.4) is fitted. This may be a separate unit connected into the system as shown in the Figure, alternatively it can be built into the receiver. It is possible for a single design of decoder to cater for both normal and encrypted channels, the normal pass straight through but the encrypted are recognized and decoded. Decoders are of course, only usable by viewers who have paid up. A system of renting them may be used so that the rent includes the charges for the programmes or alternatively it could be "pay as you view". They are unlikely to be by-passed by the clever electronics enthusiast, although many will have a go. The state of channels (i.e. encrypted or not) is usually published in satellite magazines (Appendix 1, Section A1.1).

Until encryption is firmly established, UK viewers who receive certain channels are asked to pay an annual sum for the service (Appendix 1, Section A1.3).

In addition to paying for programmes, in the UK a once and for all TVRO licence is required (Appendix 1, Section A1.3).

*1.3.2 **Regulatory***

Planning Permission: In the UK, guide-lines are issued to local authorities, but mainly concerning DBS. In essence:

(i) no permission is needed provided that the antenna is not greater than 90 cm diameter and is the only one on the premises;

(ii) the antenna must not project above the highest part of the roof.

For dishes over 90 cm, permission may or may not be needed according to how the particular local authority interprets the present guide-lines. It is prudent therefore to contact the authority in advance – there may be nothing to worry about but better safe than sorry.

Regulating Organizations: A list is contained in Appendix 7.

Chapter 2

INITIATION

As explained in the Preface, the aim of the book is to give non-technical readers a well-founded insight into the whole system with the technically minded able to expand on this via the various formulae-laden Appendices. In trying to understand how such a complex system works it is essential that some basic scientific disciplines are studied first however insignificant our technical aspirations may be. What this Chapter hopes to achieve therefore is the initiation of the reader into the mysteries of communication and indeed for many into some of the mysteries of the Earth itself. No cause for alarm, nothing is allowed to get too complicated, but becoming aware of the basics early will help in the flow of what follows.

For the moment, let us return to school:

10^2 (10 squared)
= 10 x 10 = 100

10^3 (10 cubed)
= 10 x 10 x 10 = 1,000

10^4 (10 to the fourth)
= 10 x 10 x 10 x 10 = 10,000

10^6 (10 to the sixth)
= $10^3 \times 10^3$ = 1,000,000 (one million)

the small, raised figure is called the *exponent* which is equal to the number of noughts in the original figure.

Also, shifting a decimal point one place to the left divides a number by 10 and one place to the right multiplies it by 10, e.g.

2.87 multiplied by 100 becomes 287.0,

usually written as 287.

2.87 divided by 10 becomes 0.287,

the 0 being added to confirm that this is a fractional number (less than 1).

2.1 Scientific Notation

In many branches of physics and engineering loom large strings of numbers. As a single example, the "weight" of the Earth is about

5,977,000,000,000,000,000,000 tonnes.

Figures like this are difficult to handle and there is ever the risk of getting the number of noughts wrong. Moreover many computers and calculators are unable to display such lengthy numbers so as these devices invade our homes the pressure on us to understand *scientific notation* grows. It is simply a way of expressing unmanageable numbers in shorthand form.

Firstly the decimal point is shifted so that only one numeral is on its left (normally the decimal point at the right-hand end of a number is not shown). So that the value of the number is unchanged, it is then multiplied by a multiple of 10 expressed by an exponent according to the number of places the decimal point has been moved. The above formidable number therefore reduces to:

5.977×10^{21}

Other examples are:

2,000,000 becomes 2×10^6 (strictly 2.0×10^6)

280,000,000 becomes 2.8×10^8

and to regain the original number from the scientific notation of it, simply move the decimal point the same number of places as the exponent of 10, to the right if the exponent is positive, to the left if negative. Where no numerals exist, 0's are added as required.

2.2 The Metric System

Although scientists and some countries have long used the metric system because of its inherent simplicity, many have been slow to catch up. We still wallow in a plethora of different units with different multiples (e.g. *12* inches to the foot, *3* feet to the yard, etc.). A mile still means more to some than one kilometre, pints ring more bells than litres. However on entering the electrical engineers' world the metric system must be adopted completely. The system has the outstanding advantage of using multiples, all based on 10.

This fits in well with decimals and scientific notation.

Each multiple has its own name and it is applied as a prefix to a basic unit representing for example, length (metre), mass (gram – at this stage take *mass* as being synonymous with *weight*). The prefixes which are important here are:

micro = one millionth ($\times 10^{-6}$)
abbreviated to μ (Greek *mu*)

milli = one thousandth ($\times 10^{-3}$)
abbreviated to m

centi = one hundredth ($\times 10^{-2}$)
abbreviated to c

kilo = one thousand (× 10^3)
abbreviated to k

mega = one million (× 10^6)
abbreviated to M

giga = one thousand million (× 10^9)
abbreviated to G (usually pronounced as in *giggle*).

From this, one kilogram is equal to 1000 grams (about 2.2 pounds) and 1 kilometre is equal to 1000 metres (about 5/8 miles).

One exception to the system is the metric *tonne* which does not have a prefix. It is equal to 1000 kilograms (roughly the same as the Imperial *ton*). Expressed in the normal way a tonne is equal to 1 megagram (1 Mg).

2.3 SI Units

The basic metric system as described briefly in Section 2.2 has been updated internationally into a more complete form covering all scientific and engineering disciplines and known as the "International System of Units" (Le Systeme International d'Unités, SI). We do not examine this in depth for it embraces many units. Those used are introduced as the book progresses. The prefixes are those of the metric system. As a single example, the basic unit of time is the *second* (abbreviated to s), so one thousandth of one second is written in the SI as 1 millisecond, abbreviated to 1 ms.

2.4 The Earth

In this great cosmos, the Earth is but an insignificant speck. Not to us however who live here for by the very nature of things, our lives depend on it. Small wonder therefore that so much time has been absorbed in making a multitude of measurements on this our dwelling place. Some of these measurements are essential in the study of satellites. Next, back to school again.

The Earth as an astronaut passing over London might see it is sketched in outline in Figure 2.1. For navigation and so that any place can be pin-pointed, lines of *latitude* and *longitude* are superimposed on maps. Longitude is reckoned from the *meridian* passing through Greenwich. It is measured in degrees East or West as shown in the Figure at 20° intervals. Latitude is reckoned from the equator which circles the Earth equidistant from the North and South poles. Lines of latitude at 20° intervals are also shown in the Figure. Cutting the Earth longitudinally therefore would produce wedge-shaped segments, cutting it through latitudes makes circular slices. The position of any point on the Earth's surface is fully determined by quoting its latitude and longitude. These two measurements are important for the alignment of dishes (Section 8.1.2).

Geostationary satellites are said to travel in the *plane* of the equator. "Plane" may need a little explanation. If an orange is used to represent the Earth and is cut in half, then the edge of the circular cut represents the equator and the newly exposed surface its plane. If one half of the orange is placed, cut surface downwards on a flat plate, the plate can be said to represent an extension of the plane of the equator. This concept may help in explanations but of course in space there is nothing. At the equator, the equatorial plane is simply the thin slice of space vertically above.

The familiar globe which is used to represent the Earth invariably shows the latter with the line joining North and South poles tilted at some 23 degrees. This is how the Earth is positioned relative to the plane containing both Earth and Sun. Happily for us, we can forget about this right from the start because it does not affect our considerations. If we imagine the Earth to be turned into any other position, the equatorial plane and the satellites move together with it, hence all drawings can follow the pattern of Figure 1.1, i.e. with the N-S line as vertical.

2.4.1 Dimensions

The Earth is not a perfect sphere, it is slightly flattened at the poles and bulges at the equator for which a special term *geoid* (Earth-shaped) is used. Various measurements are published with only minor differences between them. We use:

radius of Earth at equator = 6378 km (kilometres)

radius of Earth at poles = 6355 km

mass of Earth = 5.974 × 10^{24} kg (kilograms).

The difference between mass and weight is established in the next Section.

2.4.2 Gravity

For an understanding of satellite technology it is essential to appreciate, even if only superficially, certain features of everyday life which we normally take for granted. Hard to believe, perhaps but we all have inertia, not necessarily the human kind but the scientific. This is the resistance of any body to being pushed around; if moving it resists being accelerated, or slowed, if at rest it resists being moved. Inertia is proportional to mass or weight so we find that it is easier to move an ant than an elephant. It is one of those certain somethings which can perhaps be best described as a character or quality. There is nothing tangible about it and in fact it is undetected by any of the other senses. We are about to meet many such somethings and *gravity* is another of them. This is one of the most powerful influences in our lives, a force of good without which we would fall off the Earth, yet an unfriendly one to anybody slipping off a ladder. Gravity must be accepted as one of Nature's

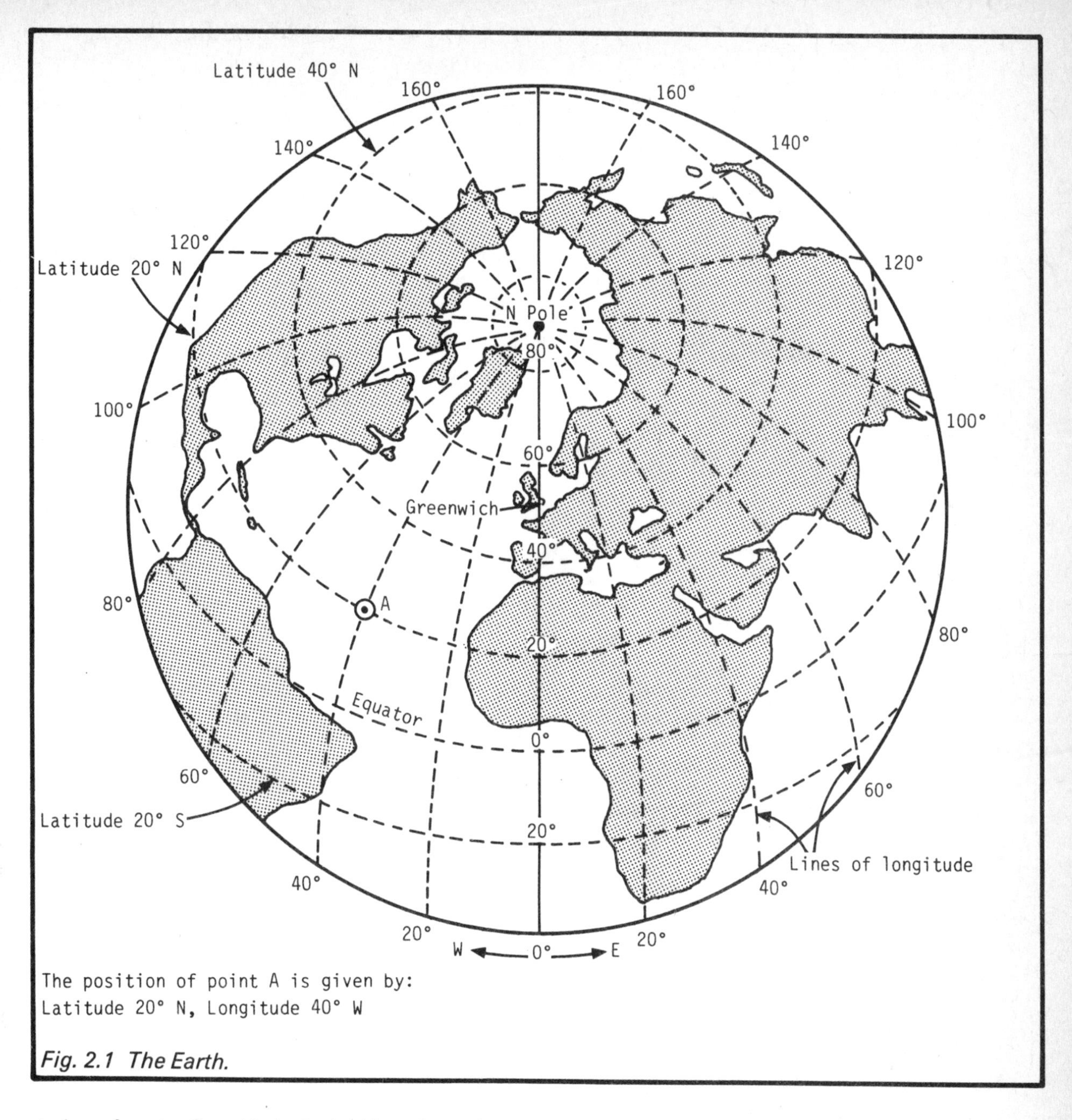

Fig. 2.1 The Earth.

strokes of genius for with it she is able to keep the Earth and all other planets in their appointed places. Although gravity itself is a mystery, its effects are well understood.

Gravity is an ever present example of the phenomenon of the force which exists between any two bodies, trying to pull them together. This force is directly proportional to the masses of the objects but inversely proportional to the square of the distance between them.[A2(1)] We go around unaware of the force of attraction other objects exert on us because compared with the Earth, all have insignificant mass. For example, a parachutist on his or her way down attracts the Earth with as great a force as the Earth exerts back. The force eventually ensures that the two meet but we know which of them moves, the one with the lower inertia.

The *weight* of a body is measured by the gravitational force with which it is attracted towards the centre of the Earth. This leads to trouble when away from the Earth's surface because the force decreases as the distance from the Earth increases. A splendid example of this is the picture of astronauts who are floating or "weightless". Hence in dealing with items up in space we cannot talk about weight so a unit which is not affected by distance from Earth is required and this is called *mass.* It represents the *quantity of matter* in a body and is in fact a measurement of its inertia. Mass is usually quoted in kilograms.

Because the Earth's gravity effectively emanates from the centre of the Earth and because the radius of the Earth varies from equator to poles, we can sum up by saying that the weight of a body varies with both latitude and altitude but its mass never varies. Throughout the remainder of the book we consider "mass" almost exclusively.

2.4.3 Fields

While on the subject of nebulous quantities, we might pause to consider what a *field* is for so often the term "field of gravity" is heard. The word is somewhat elusive of exact definition for it represents something formless, being simply another of Nature's invisibles and intangibles. On Earth we live within the field of gravity without always realizing what enormous power lurks therein. But jump off a cliff and see! From this we might say that when something produces a field other somethings are affected by it. A field is therefore a sphere of influence.

The Earth also possesses a *magnetic field* and fortunately for navigators, compasses are affected by it. Radio waves also produce fields but more on these in Section 3.1.

2.4.4 The Sidereal Day

Many years ago Charles Lamb wrote "Nothing puzzles me more than time and space; and yet nothing troubles me less, as I never think about them". We cannot afford to be so lackadaisical.

Space is no longer just something up there and time assumes ever increasing importance. What is of concern is the exact time in which the Earth makes one complete revolution on its axis and it is not 24 hours as is popularly believed. It is all very complicated because, for a start, the year cannot be divided evenly into a whole number of days. Leap years are brought about by the fact that the Earth's journey round the Sun is about 365¼ days, not 365. It is in fact very slightly less than this hence correction by an extra day on each leap year is too much so now and again the leap year is omitted (as it was in 1900). However 365¼ is sufficiently accurate for our calculations.

Not only is the Earth rotating on its axis but in one year it rotates once around the Sun and as far as the latter is concerned, the Earth rotates 366¼ times. But the (solar) year contains 365¼ days, each of 24 hours so the actual time for a single revolution of the Earth on its axis is:

$$\frac{365.25}{366.25} \text{ days,}$$

slightly less than 24 hours, in fact 23 hours, 56 minutes, 4.1 seconds.[A2(2)] This is known as the *sidereal* period of rotation, that is, as seen from the stars (from Latin, *sidus*, a star). This is of great importance in satellite technology because with geostationary satellites, arrangements must be made for completion of one revolution in exactly this time.

2.5 Decibels

The decibel is used in communications to ease the task of handling awkward ratios and large numbers. Although mainly an engineering unit it has already been enlisted by the general public to simplify day-to-day considerations of noise levels. As an example, the noise made by a pneumatic drill might be quoted as 118 decibels rather than as the less manageable ratio 6.31×10^{11}. Moreover this would presume that everybody understands scientific notation! Generally people find decibels meaningful even when the basic system is not fully understood.

Decibel notation works by using the logarithm of any ratio,[A2(3)] thereby reducing large numbers to smaller ones. There is the added attraction that when ratios are multiplied together, their decibel equivalents are simply added.

Quantities may also be expressed by the decibel system by using the dB value relative to a known basic one. The latter must be stated or it may, if well known, be indicated by a letter following the short form "dB".

The basic unit of *power* used in satellite engineering is the *watt* (e.g. a 60 watt lamp). The letter used is W, hence dBW indicates so many decibels *relative to* 1 watt. For example if a ratio of 2 is expressed by 3 dB, then 0 dBW = 1 watt, +3 dBW = 2 watts and −3 dBW = 0.5 watts. There are plenty of examples of the use of the system in Chapter 8. Table A2.1 shows some relationships between decibels and power ratios and examples are also given in A2.3 of estimating the ratio from any dB value.

2.6 Frequency

This term is a measure of *rate* of recurrence. The to and fro motion or vibration of a guitar or piano string can actually be seen and felt especially at the lower notes. Each complete vibration is known as a *cycle* and the number of cycles occurring in a stated time is known as the *frequency*. Section 2.3 affirms that the basic unit of time in the SI system is the second and the particular unit for frequency is the *hertz* (after Heinrich Hertz, a German physicist). One cycle completed in one second is expressed as one hertz. All music is due to vibration of, for example, the reed in a clarinet, string of a violin or column of air in a flute. Music frequencies vary from just below 30 hertz to a little over 10,000 hertz.

The term "hertz" is usually abbreviated to Hz. Thus technically we might say that the note A has a *frequency of 440 Hz.* It will be seen from the next chapter that radio signals "vibrate" and that each has its own basic frequency just as musical notes do. On radio receivers these are usually indicated on the tuning dial in *kilohertz* (kHz, one kHz = 1000 Hz) or megahertz (MHz, one MHz = 10^6 Hz). Note the small k but capital M (Sect.2.2). As if these numbers are not large enough, for how can anybody really appreciate something vibrating millions of times in one second, satellites work in gigahertz (GHz, one GHz = 10^9 Hz), thousands of millions of vibrations each second. There are even higher frequencies.

Summing up, the number of times a quantity vibrates in one second is known as its frequency, measured in hertz (Hz). The multiples are kilohertz (kHz), megahertz (MHz) and gigahertz (GHz).

2.6.1 The Electromagnetic Wave Spectrum

Radio and light waves are described technically as "electromagnetic", the meaning of this awesome term will become clear in Chapter 3. At this stage it is interesting to see where satellite transmission falls in the complete range of the wave frequency spectrum. This is displayed in Figure 2.2. *Wavelength* refers to the distance between the crest of one wave and the next and again we must wait for Chapter 3 for this to make sense. Very briefly, the higher the frequency, the shorter the wavelength. The bottom part of the drawing shows another way in which radio waves are classified from which terrestrial tv is seen to be within the uhf range whereas satellite tv is at higher frequencies altogether at shf.

Again there are singers and violinists who can hold a note so that eventually a wineglass vibrates to such a degree that it shatters. Not easy because only one note held long enough is successful. The frequency of the note must be exactly that of the natural frequency of the wineglass. It might be said that the sound wave gets *into tune* with the glass. A final example of vibration at the natural frequency is given by the tuning fork, it vibrates at one frequency only for if it deviated at all its tuning days would be over.

Getting into tune is one of the fascinating concepts on which radio is based. It is this facility which enables a radio receiver to pick out one station while rejecting all others. Special electronic *tuning circuits* are designed to *resonate* to one particular frequency only and usually a circuit can be made to tune to other frequencies by rotation of a knob or pressing a button. A transistor radio is surrounded by scores of radio transmissions yet it selects the one required with comparative ease.

Resonant (or tuned) circuits are essential in all radio and tv receivers.

2.7 Power, Signals and Noise

A peculiar mixture indeed so let us first understand the terms. In everyday life, power is something

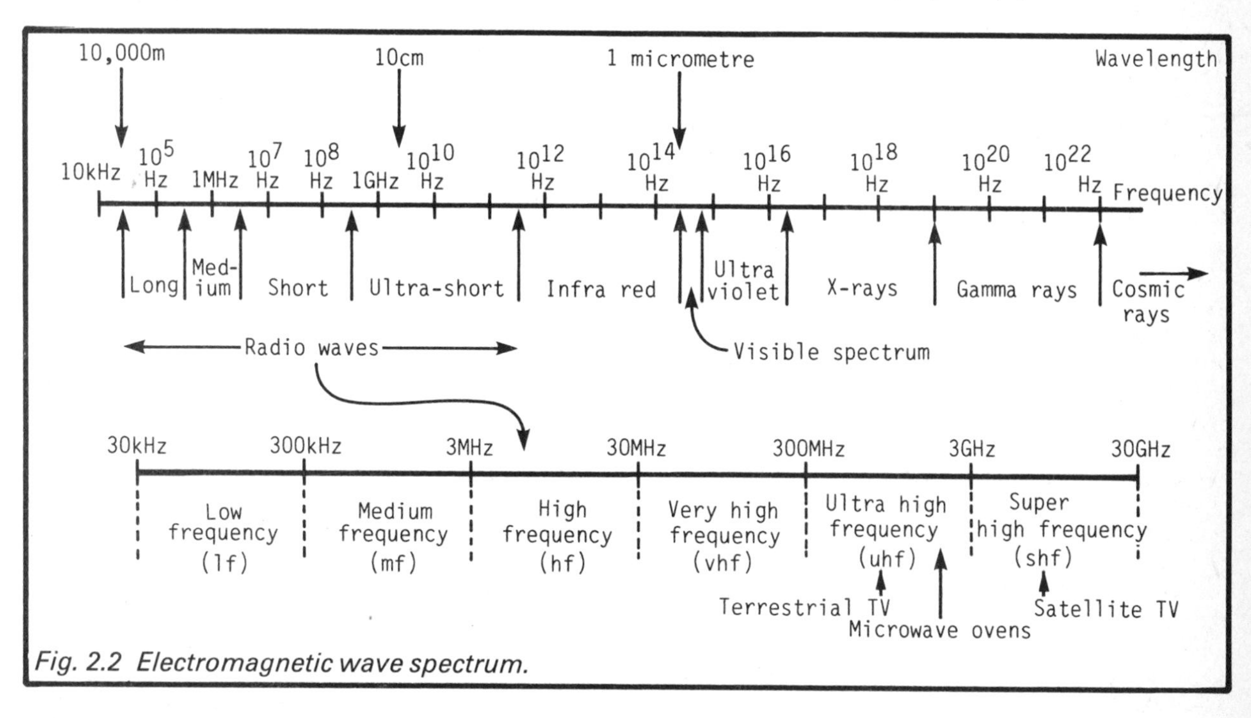

Fig. 2.2 Electromagnetic wave spectrum.

2.6.2 Tuning In

Let us go back to childhood on the playground swing. Even at a tender age we sensed how to keep things going by giving the swing pushes at the right moments. A swing has a natural rhythm or vibration frequency which depends on its length and the frequency can be calculated as it can for any pendulum. Giving the swing pushes at the right point in its cycle increases the travel and keeps it going. This is *resonance.*

engines develop, signals control traffic and noise is that which assails our eardrums. In communication engineering however these terms have extended and definable meanings.

We can look on *power* as the ability to do something although technically it is the rate at which it is done. The electricity mains have power for they can heat rooms and drive lawn mowers. What concerns us is the comparatively insignificant power possessed by a radio wave for small though it may be, it excites the atomic particles in an antenna to lead eventually to a tv picture. Just to get powers into some perspective, that in a radio wave can be as little or even less than one million millionth (10^{-12}) of that of a one-bar (1 kilowatt) electric fire. It is measured in the same unit, the *watt* so whereas the fire *dissipates* a power of 10^3W, the radio wave under discussion has a power of $10^3 \div 10^{12}$ or $10^3 \times 10^{-12} = 10^{-9}$W, i.e. one thousand millionth of a watt, infinitely small but still effective when properly handled.

A *signal* according to the dictionary is "an intelligible sign conveying information" and the term is used to embrace all electrical and radio intercommunication, from the dots and dashes of the early Morse code to the highly complex affair of colour tv. *Noise* is explained as "irregular fluctuations accompanying but not relevant to a transmitted signal". We can perhaps get to grips with these two definitions by starting with the first two people arriving at a party. They converse and there is no difficulty. A adjusts his or her voice to provide a *signal* which B hears comfortably and vice versa. More guests arrive who chatter among themselves. To either of the first comers when listening to the other there is now a background *noise* making reception difficult. Technically we say that the *signal-to-noise ratio* (i.e. loudness of signal/loudness of noise) has decreased, simply because the loudness of the noise has increased. In other words, whereas originally the voice signal had practically no noise to compete with, now it has. Of course one alternative is to tell the others to be quiet. This can hardly be recommended so an alternative procedure is adopted which is that by raising the voice, A, for example, increases the signal-to-noise (s/n) ratio and B hears comfortably again. As more guests arrive the general noise level rises aggravated by the fact that all are talking with raised voices because of the background noise. Hence the s/n ratio decreases still more, and the noise is beginning to win. In fact B's ears can receive more noise than A's signal and at this stage conversation is really difficult. Given sufficient noise, conversation is impossible even when one person shouts into the ear of another.

What this demonstrates is that the efficacy of communication is dependent on how greatly a signal exceeds any noise accompanying it. We have shown this to be so for the audio case but in fact it is a general problem throughout communication. Noise is the snake in the grass. To the communications engineer noise is not limited to only that which we hear, any unwanted electrical "irregular fluctuation" is called noise, be it audible or not. As a practical experiment, remove the antenna plug from a tv receiver and the screen is immediately filled with white specks. These are due to electrical "noise" generated within the set. A tv picture is spoilt if the incoming signal is not sufficiently strong compared with this internally generated noise.

To conclude this Section, we can put a few very approximate figures to the noisy party. With a s/n ratio of 16 (signal 16 times louder than the noise), all is well. When the s/n ratio drops to 1, there is difficulty and words get lost (fortunately ears are adept at filling in). However, when the ratio falls to say, $\frac{1}{16}$ (which is very poor) there is much trouble and little is understood. Working now in decibels for experience we start off with a s/n ratio of 16 which according to Appendix 2[A2(3)] is equivalent to +12 decibels (dB). A ratio of 1 is equivalent to 0 dB and a ratio of $\frac{1}{16}$ to −12 dB. When the signal power is greater than the noise power therefore, the s/n ratio expressed in decibels is positive, when the signal power is less, the ratio is negative.

Note the use of the word "power". We work mainly in terms of this; "loudness" as considered in the noisy party has meaning only when ears are present.

2.8 The Electron

It is not possible to interest ourselves in a subject such as this without having some feeling for the key to it all, the *electron.* But first let us look at its parentage, the *atom.* There are about one hundred different atoms around and they are the particles which hang together in their millions and millions to make up many of the substances of daily life. Take copper for instance, pure copper consists of copper atoms tightly bound together and nothing else. For this reason it is known as an *element.* Pure iron is a different element, it has only iron atoms in it. Oxygen and nitrogen, the main constituents of air are also elements. However when some elements get together things can happen and *compounds* are formed. One well known example is water. By mixing hydrogen and oxygen gases in a vessel and exploding them to get things going, there remains only a tiny drop of water, of very different form from the constituent gases. Water is therefore the child of two parents and is not an element but a compound made up of *molecules*, each of two atoms of hydrogen linked with one of oxygen, hence the chemist's H_2O.

Hard to believe is the fact that every atom is like a miniature solar system. It is well ordered, not like the higgledy-piggledy group of planets we live in. The atom has a central *nucleus* like the Solar System has the Sun and round the nucleus moving in orbits (more on these in Chapter 5) at tremendous speeds are its electrons. They travel round several thousand million times in one millionth of a second! Figure 2.3 pictures a copper atom with its electrons shown as little balls but this is only because we have no idea what they really look like for they are too small ever to be seen. Insignificant though they may be, they do have some mass (Sect.2.4.2), so small as to be beyond comprehension except via scientific notation. The mass of the electron is 9.1×10^{-28} grams (one gram is about one thirtieth of an ounce).

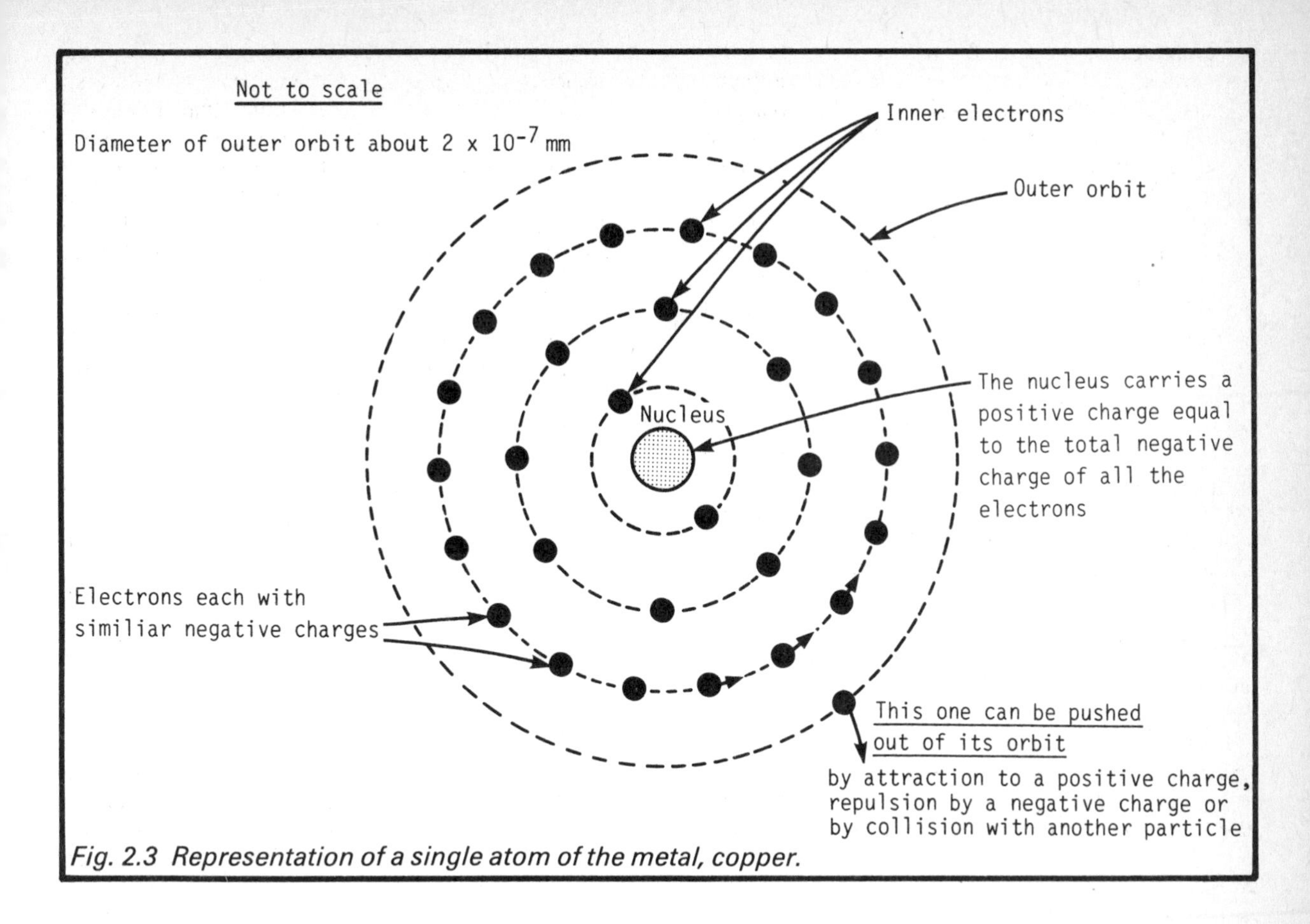

Fig. 2.3 Representation of a single atom of the metal, copper.

2.8.1 Positive and Negative
So much for the almost unbelievable facts about the electron. What is of greater interest is how they behave towards one another. Every electron carries, or rather *is*, a minute *charge*, the stuff electricity is made of. Section 2.4.3 brings in the subject of *fields* and charges have fields surrounding them just as the field of gravity surrounds the Earth. The electron charge is labelled "negative" and it has an opposite number, called "positive". We must not try to understand what a charge is, like gravity it is something Nature has bestowed upon us. Every electron carries *exactly* the same amount of charge.

The golden rule about charges is that

"like charges repel each other, unlike attract".

From this it is clear that electrons are not too friendly towards their neighbours but are themselves welcomed by a positive charge. This is how the atom is held together for the nucleus has a strong positive charge sufficient to keep all its electrons in their appointed places.

In some materials such as silver and copper, some electrons are quite easily pushed out of their atom orbits (see Figure 2.3), leaving them free to move in the spaces between the atoms. Normally the positive and negative charges of the atom balance exactly so if an atom loses an electron it becomes positive and it is then known as an *ion*. If therefore a negative charge is connected to one end of a piece of copper wire and a positive charge to the other end (e.g. from the two terminals of a battery), then free electrons will flow away from the negative end (like charges repel) towards the positive end (unlike attract) and hey presto! there is an electric current.

A sketchy explanation indeed but it should be of some help when we tackle radio transmission, remembering always that charges give rise to fields which move electrons according to the rule above and when electrons move en masse, this constitutes an electric current.

Chapter 3

THE RADIO WAVE

It was in the 1880's that Heinrich Hertz conducted his famous experiment confirming that communication by radio was possible. Each time he generated an electric spark, a second smaller one could be produced by equipment several metres away – and there were no wires between them. Some of the energy generated by the first spark had travelled across the room in the form of radio waves to be picked up by the second circuit. "Wireless" had arrived. In only 100 years the technology has exploded, from distances of a few metres to many millions and from the simple spark to the whole gamut of data and tv we know today. In fact as we read this the air around us is alive with the fields of a thousand and one different radio transmissions. Unbelievable? – switch on a portable radio and it will identify many of them one by one.

Nobody will pretend that fully understanding the radio wave is child's play, it is not. Accordingly any skirmish with it must of necessity be superficial. We make a start by assuming a wave which is carrying no information. It is then said to be *unmodulated* as opposed to the normal wave which carries information and hence is *modulated.* Modulation is discussed in Section 4.4.

3.1 A Split Personality

The experiments carried out by Hertz arose from his interest in the theories of James Clerk Maxwell, a Scottish physicist. Maxwell had earlier managed to associate what was then known of electricity and magnetism with the laws governing the behaviour of light. From this emerged the startling conclusion that electricity and magnetism together *are* light, not only light but they are also the ingredients of all radio waves. What distinguishes one wave from another is the frequency (Sect. 2.6) and light is characterized as a collection of waves at very high frequencies indeed but which have the unique capacity for stimulating the sense of sight.

A radio wave is described as *electromagnetic*, having two parts, one electric and the other magnetic. The two parts are in the form of fast moving fields (Sect. 2.4.3). An electric field reveals itself in the home quite openly. Switch on the tv set and place a small piece of paper or card very near to or on the screen, it remains held in position by the electric field there. Place the back of the hand close to the screen and the effect of the field can be felt on the skin. It is a hair-raising event. Also the screen glass needs to be cleaned more frequently than normal because the electric field there attracts particles of dust from the air. Pull off certain synthetic garments and when held up they fly towards us, an electric field is responsible. A magnetic field will not do any of these things but it has its own ideas. The Earth's magnetic field provides an example, it is around us everywhere yet we see nothing of it. It shows itself by its action on a compass needle which turns to indicate where the magnetic poles are. Place a magnet near a pin, the strong magnetic field is there and the pin reacts to it.

It is evident that fields somehow contain *energy* for they are capable of doing work and this is how a radio wave *excites* a piece of wire or an antenna. The electric field is capable of moving *electrons* (Sect. 2.8) in the wire to create an electric current.

Now comes the difficult part – the electric and magnetic fields are at right angles to each other and to the direction of propagation. This implies that the fields themselves have *direction* which we know is true from the compass. If it is remotely possible to imagine two different fields linked together and approaching us, then if the electric field varies in strength from side to side with the magnetic field from top to bottom, the wave is said to be *horizontally polarized.* If the other way round, it is *vertically polarized*, so it is the electric field which counts. We might imagine an oncoming wave as a sheet of graph paper on which the sets of lines represent the two fields. It is essential to appreciate this because both forms of polarization are used with satellites. They are usually denoted by V and H.

3.1.1 Polarization

The dictionary defines *imagination* as a "mental faculty forming images" and this is certainly an essential requirement of this Section. Seeing in the mind a radio wave, travelling at fantastic speed and at the same time vibrating thousands of millions of times each second is demanding enough but as indicated in Section 3.1 we have to visualize in addition the two fields and how they interact.

Normally the electric and magnetic fields are at right angles to each other and they lie in a plane transverse to the direction of propagation. If they maintain their relative directions, e.g. the electric field always vertical, then the wave is said to be *linearly* polarized. Hence for any particular satellite channel, not only must the frequency be quoted for tuning in, but also the polarization, V or H. This is essential because for maximum signal pick-up the polarizations of the transmitting and receiving antennas must match. We see this with terrestrial tv antennas which have either vertical or horizontal rods, a clear indication of the polarization in use. For satellites the dish must also be arranged accordingly. Unfortunately when this is done it is unable to receive channels on the opposite polarization.

There are ways of getting over this as discussed later.

One bonus is that the same frequency can be used for completely different transmissions provided that they are on opposite polarizations, although just to be on the safe side they are usually kept well apart.

What complicates the issue further is the fact that waves from DBS satellites have a different kind of polarization. In its passage through the Earth's atmosphere it is possible for the polarization of a microwave to become twisted so that it is no longer truly vertical or horizontal. Unless the receiving antenna is rotated accordingly, there is a reduction in signal pick-up. The effect is known as *depolarization.* However if *circular* polarization is used, depolarization has no effect on the received signal. In this case the electric and magnetic fields remain at right angles to each other but now the pair continually rotate. It is as if a cowboy were moving towards us with his lasso whirling in a circle. Both rope and wave can rotate in either direction and if we look towards him or the transmitting antenna, then a *clockwise* rotation is described as *right-handed* and *anti-clockwise* as *left-handed.*

We need not, nor indeed can we, go into the finer points of polarization (it all gets terribly mathematical), it is only necessary to remember that there are four different polarizations. Generally the earlier satellites employ linear polarization and are therefore labelled V or H. DBS satellites use circular polarization for which the labels are LHC and RHC (left- and right-hand circular).

3.2 Wavelength

Whatever the frequency of a radio wave its speed of propagation in a vacuum and generally in air too, is 3×10^8 metres per second (m/s) – always. There's food for thought; the waves do not start from zero and accelerate to their final speed as everything else which moves on Earth does! The speed is denoted by the single letter c and is equivalent to 186,000 miles (7 times round the Earth) per second. Nothing in the universe goes faster. Nevertheless with space travel the distances are such that there is a delay of over one second from the Moon and even the "live by satellite" programme arrives one-quarter of a second late.

If a wave has a frequency denoted by f Hz, then in one second there will be f vibrations or cycles (the latter term is preferred) spread over c metres. The length in the air or space therefore of one cycle is c/f metres and this is known as the *wavelength.* If we could achieve the impossible by stepping aside to view a wave as it passes and then plot a graph of its strength over the time for one wavelength to pass, the resulting graph would have the shape so familiar to all engineers, a *sine* curve as shown in Figure 3.1. The Figure shows as best it can how, for example, the electric field varies, changing over from one direction to the other in the middle of the cycle.

A radio wave can therefore be known by its frequency or its wavelength and just to confuse us, either term is used. But we can easily convert from one to the other, for, denoting the wavelength by

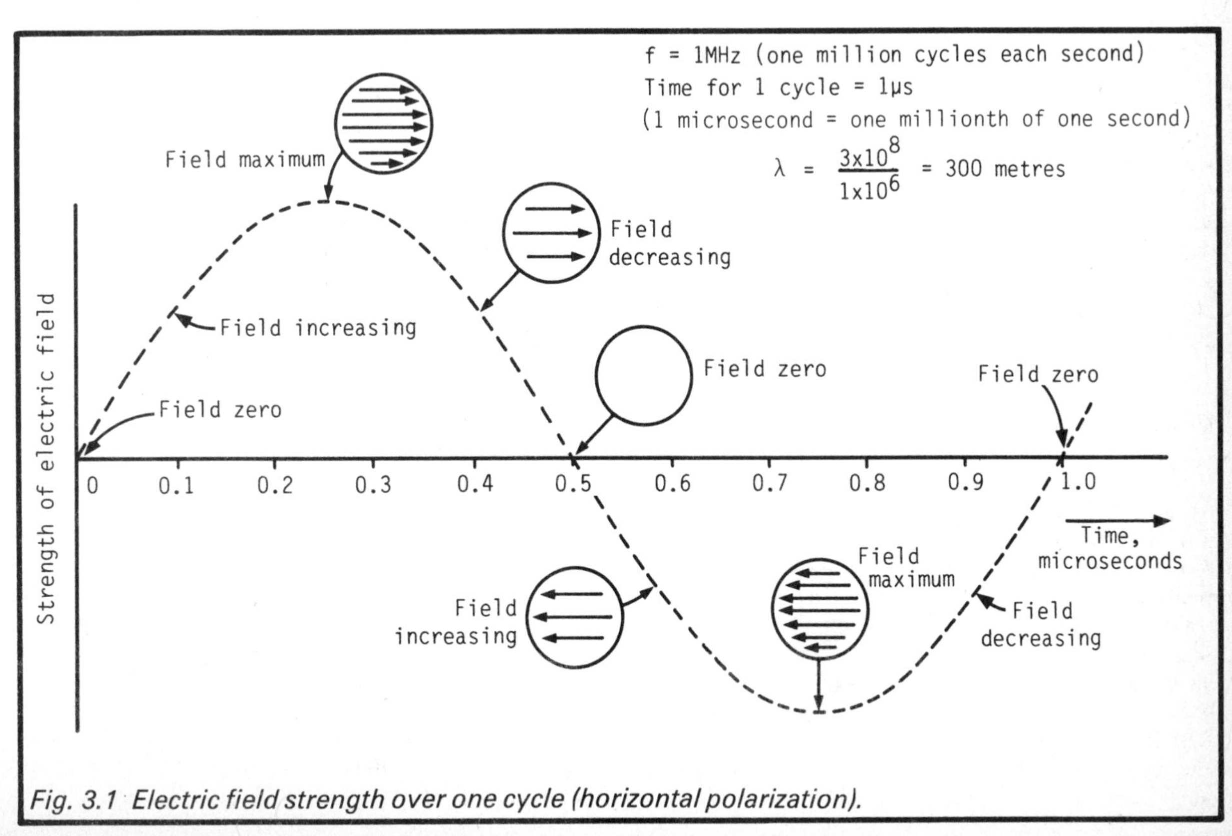

Fig. 3.1 Electric field strength over one cycle (horizontal polarization).

λ (the Greek, *lambda*):

$$f = \frac{c}{\lambda} \quad \text{and} \quad \lambda = \frac{c}{f}$$

so the radio wave from a station broadcasting on a frequency of 600 kHz has a wavelength, λ of

$$\frac{3 \times 10^8}{600 \times 10^3} = 500 \text{ metres.}$$

Compared with this the wavelengths used with satellites are very short indeed, around 2.5 centimetres.

3.3 Propagation

When a radio wave leaves a transmitting antenna several factors control the strength it will have on reaching some distant point, the power it starts off with, its frequency, the distance travelled and the losses it suffers on the way. Although our main studies are in the gigahertz region, it is worthwhile discussing waves of lower frequencies first to understand why satellites use only the very high frequencies which are notoriously difficult to handle.

The wave on its journey has to contend with either absorption by the ground or reflections and absorptions in the atmosphere, or both. Figure 2.2 is useful here as a guide. At low frequencies antennas transmit the wave over the surface of the Earth so because of absorption of energy from the wave by the Earth and solid objects its strength diminishes with distance of travel. Moreover the losses increase with frequency so that whereas at 200 kHz the signal power 100 km away may be ample for radio broadcasting, for the same transmitted power, at a frequency of 3 MHz it could be reduced to as little as one five-hundredth of this. Clearly ground wave transmission has its limitations and is unsatisfactory for long distances as frequencies increase above a few megahertz.

Radio broadcasting at higher frequencies therefore avoids using the ground wave as much as possible. In fact it makes use of a peculiar condition high up in the atmosphere. Between about 100 and 1,000 km up and running completely around the Earth, are layers of *ionized* air, forming what is known as the *ionosphere.* Here radiation from the Sun causes electrons to break away from their atoms (Sect. 2.8.1), the atoms then become positive ions and have the effect of bending a radio wave back to Earth as though the atmosphere there were a radio mirror. Because the radio waves are sent upwards through the atmosphere they suffer less loss than when travelling over the ground, accordingly long distance communication is possible as any short-wave enthusiast knows.

Unfortunately this useful facility of reflection back to Earth by the ionosphere begins to fail above about 100 MHz for the mirror effect decreases as wave frequency rises and eventually disappears altogether above about 2 GHz. Waves at higher frequencies than this therefore shoot straight through the ionosphere and are normally lost. The ionosphere also is notoriously fickle as the various layers shift about. Hence because of this and the fact that the lowest frequencies suitable for terrestrial tv are in the uhf range, the ionosphere is of no help so we are back to the ground wave again. The ground losses make transmission satisfactory over short distances only which is why many transmitting stations are required to service a comparatively small area.

Now it is possible to understand why satellites work at even higher frequencies. By working at a few gigahertz the wave can be projected straight through the ionosphere into space and back again with only a little interference through absorption and rain attenuation on the way. Compared with a ground wave therefore the satellite wave suffers little loss with distance travelled. Moreover through space above the ionosphere there is no loss at all hence transmission over very long distances is possible. By avoiding ground losses therefore, one transmitting station plus one satellite can serve an area which otherwise would require hundreds of terrestrial stations.

3.4 Frequencies for Satellite Working

Section 3.3 shows that satellite frequencies must be higher than those which are turned back by the ionosphere so that they pass straight through. Then Section 2.7 gives reasons why the signal-to-noise ratio must be adequate for successful communication or tv picture. One of the difficulties of space communication is that receiving antennas, especially home ones, are unable to avoid "seeing" regions of space around the satellite they are aimed at. From these regions arrive various kinds of radio noise which are especially harmful when the signal itself is weak. This "sky noise" is at its lowest between about 1 and 10 GHz, above this it gets worse up to about 20 GHz. For these reasons and those given in the previous Section, satellite frequencies can only be in the low gigahertz region.

It is clear that countries simply cannot "go it alone" otherwise their transmissions would interfere with others and in the resulting chaos nobody would win. Accordingly people got together and at the World Administrative Radio Conference (WARC) in 1977 the various bands of frequencies for the more powerful DBS satellites were allocated. Europe got 11.7 to 12.5 GHz which is sufficient for 40 separate tv services from each of 4 satellites.

Apart from the effects of sky noise, many other technical features combine to dictate the choice of frequency bands for satellite working but two are especially worthy of mention:

(i) we will see in Chapter 8 that the amount of signal a dish can pick up increases with both frequency and dish diameter.A8(15) For a given signal pick-up therefore one can be traded for the other, e.g. by using higher frequencies to allow the use of smaller receiving antennas.

(ii) microwave ovens do their job by employing microwaves (see Figure 2.2) to excite the molecules of water in food. Their extra vibrations produce heat and the power for this comes from one source only, the wave. Such action is desirable – it cooks our food. However when communication microwaves travel through the atmosphere and meet water in the form of rain or mist the same action is there but now undesirable because power is extracted from the waves. Such a loss of power can be ill afforded. The extent of the action depends on many factors especially the relationship between water droplet size and transmission wavelength but overall the losses are greater above the frequency range chosen.

Summing up, the chosen range allows for comparatively small receiving antennas while at the same time avoiding the worst of the rainfall losses which occur at even higher frequencies.

Since this allocation, progress in satellite and receiving equipment technology has enabled other "medium powered" systems to be developed of almost the same overall performance as originally envisaged. These are on different frequencies of course, e.g. the satellite ASTRA from the Société Européene des Satellites downlinked on 11.2 to 11.45 GHz, but again within the same range of "minimum sky noise" frequencies.

The channels for the European DBS satellites are labelled 1 to 40 and the frequency of any particular one can, for those with their calculators handy, be obtained from:

$$11.7083 + (0.01918 \times \text{Channel Number})\ \text{GHz}.$$

The UK has been allocated Channels 4, 8, 12, 16 and 20 from satellite position 31°W (RHC polarization – Section 3.1.1).

Spain gets Channels 23, 27, 31, 35, 39 from the same satellite but with LHC polarization, and the Netherlands have the same channel numbers as Spain but with RHC polarization and from a different satellite position at 19°W.

This is an example of how the same frequencies can be employed more than once provided that the footprints are well separated geographically and different polarizations are used. Spain and the Netherlands are not exactly neighbours and have different polarizations hence the required degree of

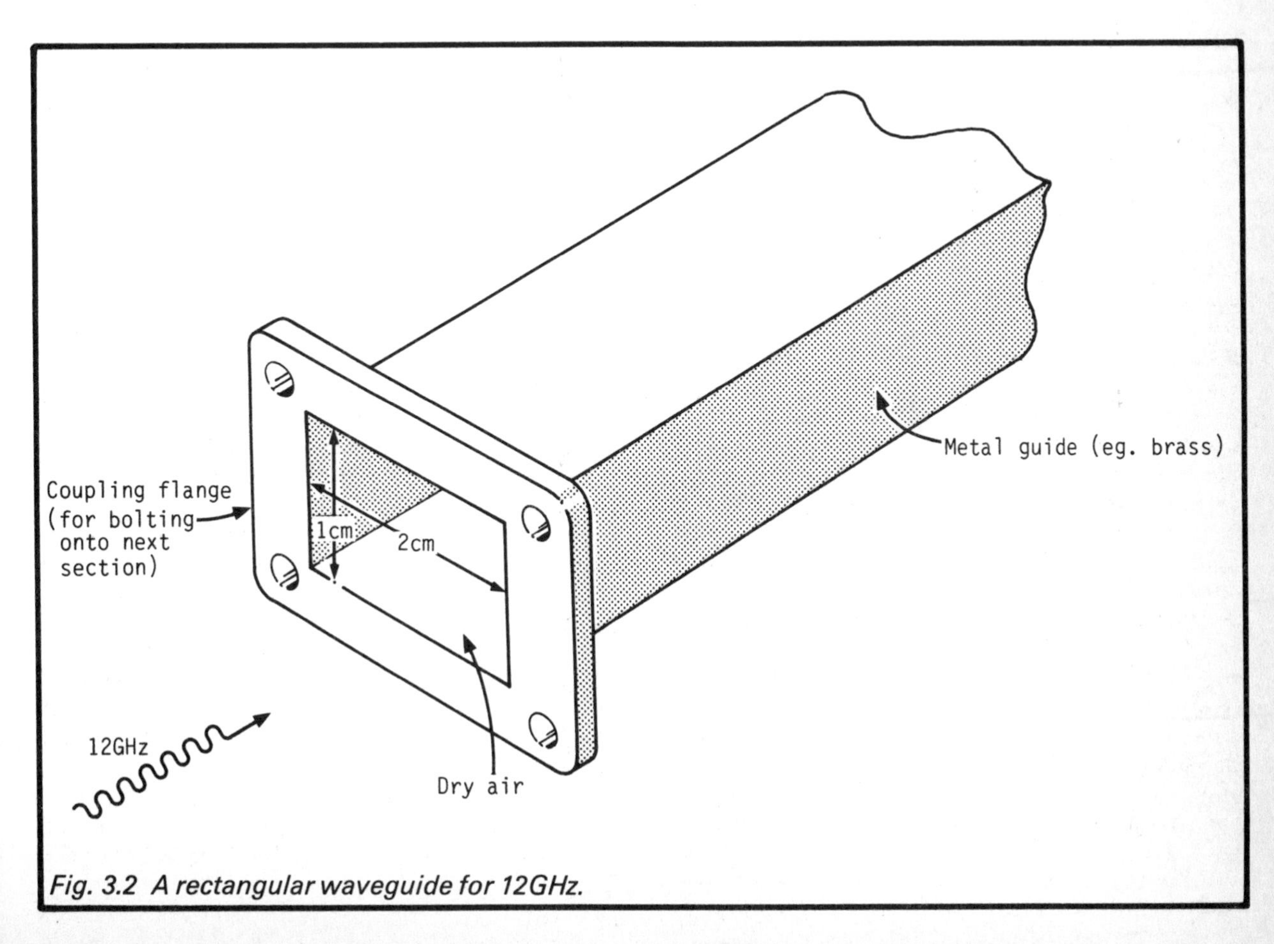

Fig. 3.2 A rectangular waveguide for 12GHz.

freedom from interference between any two channels operating on the same frequency is obtained.

A user need have no fear of this conglomeration of frequencies, polarization and channel numbers for on most satellite receivers the programme is simply selected by its channel number and features such as automatic polarity selection may also be built in.

A complete list of the satellites for Europe and the channels allocated to each country is given in Appendix 3.

3.5 Waveguides

As wave frequencies get higher so the problems associated with transmitting them over wires or cables increase. A pair of wires in any shape or form is satisfactory for telephone connexions whereas for tv a special type of cable (see Section 9.2.5), known as *coaxial* is employed otherwise the losses become excessive. At higher frequencies still, say above 0.5 GHz it becomes possible to direct or guide radio waves along special hollow tubes usually of metal and of rectangular or circular construction. The losses in these *waveguides* are only a fraction of what they would be over a coaxial cable. Usually the electric and magnetic fields are at right angles to each other and perpendicular to the direction of propagation (which is along the tube). This is the same condition as is found for the plane radiated wave (Sect. 3.1).

There is a close relationship between the internal dimensions of the guide and radio wavelength. Typically the dimensions might be 2 x 1 cm to carry a 12 GHz transmission which has a wavelength of 2.5 cm (Sect. 3.2). An example is shown in Figure 3.2. The guide is sealed and filled with dry air or gas to avoid absorption of the waves by moisture. In a satellite a waveguide similar to the one shown in the Figure carries the radio wave to the transmitting antenna (Sect. 6.2.1).

Chapter 4

SIGNAL PROCESSING

Much in the electronic world of today is accomplished by manipulating waveforms and signals to make them perform in some particular way. This is known as *signal processing.* In this Chapter we look at those basic operations without which there can be no radio or tv at all, only from the point of view as to what can be done, not so much *how* it is done. Most of the devices employed nowadays are in the form of integrated circuits, practically everything built onto a tiny chip of silicon. The electrical circuits within an integrated circuit are usually so complex that few people, except for the designers themselves, can understand them. There is hardly any point in our trying therefore.

4.1 Amplification

When an electrical signal moves through any material it meets a certain amount of resistance just as water does when flowing along a pipe. This may be through wire, fibre or the atmosphere. In overcoming the resistance, power is required and this can only come from the signal itself. Accordingly as the signal travels along it becomes progressively weaker, i.e. its amplitude or power is reduced. The process is known as *attenuation* or *loss* and it is measured most conveniently in decibels (Sect. 2.5). For example, when a signal suffers an attenuation or loss equal to 3 dB, it is reduced to half its original power because as Table A2.1 shows, (–)3 dB is equivalent to a power ratio of 0.5.

It is now clear why radio waves, although attenuated by the atmosphere, are not by space. Nothing in space offers resistance.

The effects of attenuation are counteracted by *amplification.* To amplify means to enhance or increase the strength of an electrical signal. An amplifier is an electronic device with an input for the weak signal, an output for the signal after amplification and an input for a supply of power. The electronic engineer's symbol for an amplifier is at (i) in Figure 4.1. The output of an amplifier should resemble the input in all respects except for an increase in magnitude and nothing (e.g. noise) should be added – a whisper into a shout so to speak. Amplifiers are not only used to neutralize the effects of attenuation, other duties include the production of sufficiently high-power signals for example for driving loud speakers, radio and satellite transmitters.

Amplifier gains are usually expressed in decibels, for example, an amplifier with a gain of 20 dB increases the power in a signal one hundred times as it passes through (Table A2.1).

4.2 Oscillators

These are our wave generators. Their job is to produce a waveform either "pure" in which no other frequencies are present than the intended one, or complex in which a *fundamental* frequency is mixed with higher ones. The two types might be distinguished in the audio range by the pure waveform being of the form of an unexciting and monotonous whistle whereas the complex is that of the harmonious tones of musical instruments. The pure waveform oscillator has the greater use in satellite transmission.

Frequencies range from 20–30 Hz for the pedal notes of an electronic organ to the many gigahertz of microwaves. The symbol used is shown in Figure 4.1(ii). There is no input terminal for the device generates of its own accord. Naturally a power supply is required but the symbol takes this for granted.

4.3 Bandwidth

Vibration and frequency are mentioned in Section 2.6. We ourselves are the proud possessors of built-in frequency generators – the vocal chords. By using muscles to change the tension in the chords, blowing through them and then by use in the oddest

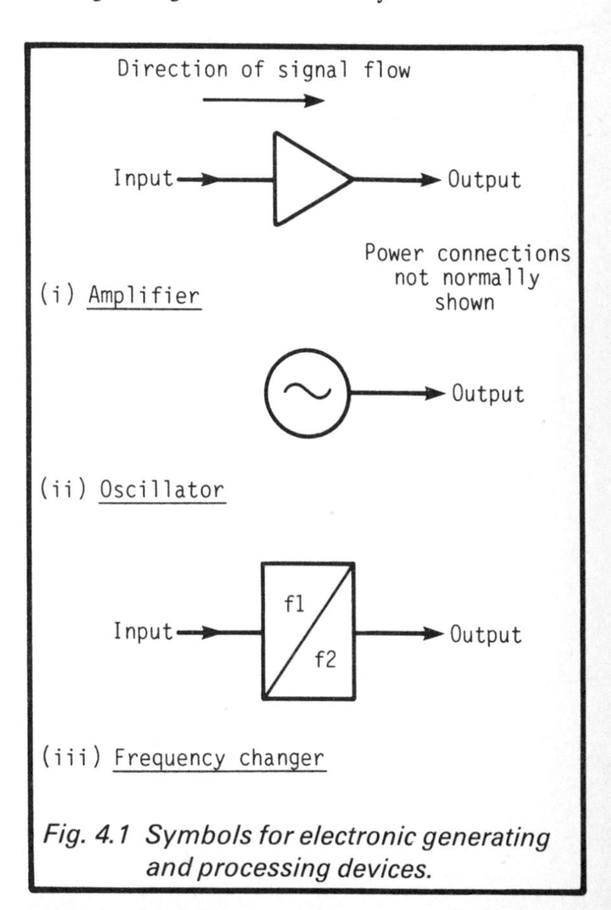

Fig. 4.1 Symbols for electronic generating and processing devices.

of ways of tongue, lips, palate and teeth, speech comes out and sometimes we actually say something of importance. If a few words were captured by a microphone and analysed to see what frequencies were present, they would probably be found to be in the range from the lowest at about 100 Hz to the highest at, say, 8000 Hz (8 kHz). Analysis of orchestral music would tell a different story, between around 30 Hz up to 10–15 kHz. In both cases the generator (speech or orchestra) is said to have a frequency *bandwidth*, which is simply the number of Hz between the lowest and highest frequencies, for speech about 7900 Hz, for high-fidelity music nearly 15 kHz.

In conversation or listening directly to music, the *channel* which carries it to a listener is open air and this conducts the sound faithfully in that no frequencies are treated differently from the rest. This may not be the case when distance is increased and an electronic channel such as a telephone or radio circuit is brought into use. In this case some frequencies may not get through at all, especially the higher ones. It is important for tv transmission that this does not happen.

Section 7.5 shows that tv pictures need about 6 MHz (6,000,000 Hz) bandwidth so clearly a channel designed for hi-fi music cannot handle tv signals, no frequency above 15 kHz would get through and the picture received would be unrecognizable. The rule with bandwidth therefore is that a quart cannot be put into a pint pot unless some degradation of the overall performance is tolerated.

4.4 Modulation

The question is how can the band of frequencies which makes up a tv picture plus its sound, be broadcast to millions of viewers? Clearly the idea of radiating these frequencies directly cannot be considered for with more than one station transmitting, viewers would receive a mixed bag of pictures and be unable to separate them.

The technique which has been employed since radio began, and in fact is the basis of all radio, is known as *modulation.* With it a higher frequency wave *carries* (on its back, so to speak) the lower tv or other signals which are known as the *baseband.* For technical reasons the frequency of the carrier wave must be many times that of the maximum baseband frequency it transports. Terrestrial tv provides an example where a carrier frequency of some 200 MHz or more carries a baseband (the tv signal) of about 6 MHz.

A modulator is a complex electronic circuit for which the symbol only is shown in Figure 4.2(i). The baseband input (the modulating frequencies) and the carrier frequency (Sect. 4.2) are fed into the modulator. These are then mixed in a special way and the modulated wave is taken from the output terminals as shown.

Imagine, if you can, a 200 kHz carrier wave and consider that it has to carry (i.e. be modulated by) a 500 Hz continuous piccolo note (about upper C on the musical scale). The simplest method by which this can be achieved is known as *amplitude modulation* (am), the method generally used for medium wave broadcasting. Unfortunately for us it is not used for satellite transmission so we have to struggle with a more complicated method known as *frequency modulation* (the well-known fm). With this the carrier wave has constant amplitude but its frequency varies. The degree of frequency variation is proportional to the amplitude of the modulating wave whereas the rate of variation is according to that of the modulating wave itself.[A4(1)] Thus the 200 kHz carrier wave might vary between 190 and 210 kHz 500 times each second when modulated. If someone blows the piccolo harder so that the amplitude of the modulating wave increases, the variation could be between, say, 180 and 220 kHz, again changing from one to the other 500 times each second. Simplified to the extreme perhaps but this is the basic principle on which fm rests. Note that the *rate* of frequency variation is controlled entirely by the frequency or frequencies of the modulating wave but that the maximum *degree* of frequency variation is set by the system designer. It is called the *deviation.* FM has the distinct advantage over am in being less affected by noise. It can achieve the same signal-to-noise ratio (Sect. 2.7) with much less transmitter power than is required by amplitude modulation, hence because satellite power is at a premium (it all has to be collected from the Sun), fm is the obvious choice.

One of the fundamentals of communication engineering is that while a frequency is being changed, other frequencies are generated, known as *harmonics.* To *demodulate* the wave later, i.e. regain the baseband, the harmonics generated in the modulation process must be present, none lost on the way.

Taking more practical examples, it can be shown[A4(3)] that the bandwidth required by an fm radio broadcast transmission is as much as 180 kHz so a station quoted as broadcasting, for example on 90 MHz in effect broadcasts over the range 89.91 to 90.09 MHz but only when the modulation has maximum amplitude. If this seems a little complicated, don't worry, it is, that is why we have looked at the process in simple terms only. In practice a multitude of different frequencies in the baseband create components of the carrier shifting in frequency at indescribable rates. In addition the changing levels of these frequencies cause deviations to vary rapidly over the whole range. Altogether a mêlée of activity, impossible to visualize. But fm works and works well.

From all this we should remember that:

(i) modulation is the technique by which a carrier wave has impressed on it a band of lower frequency

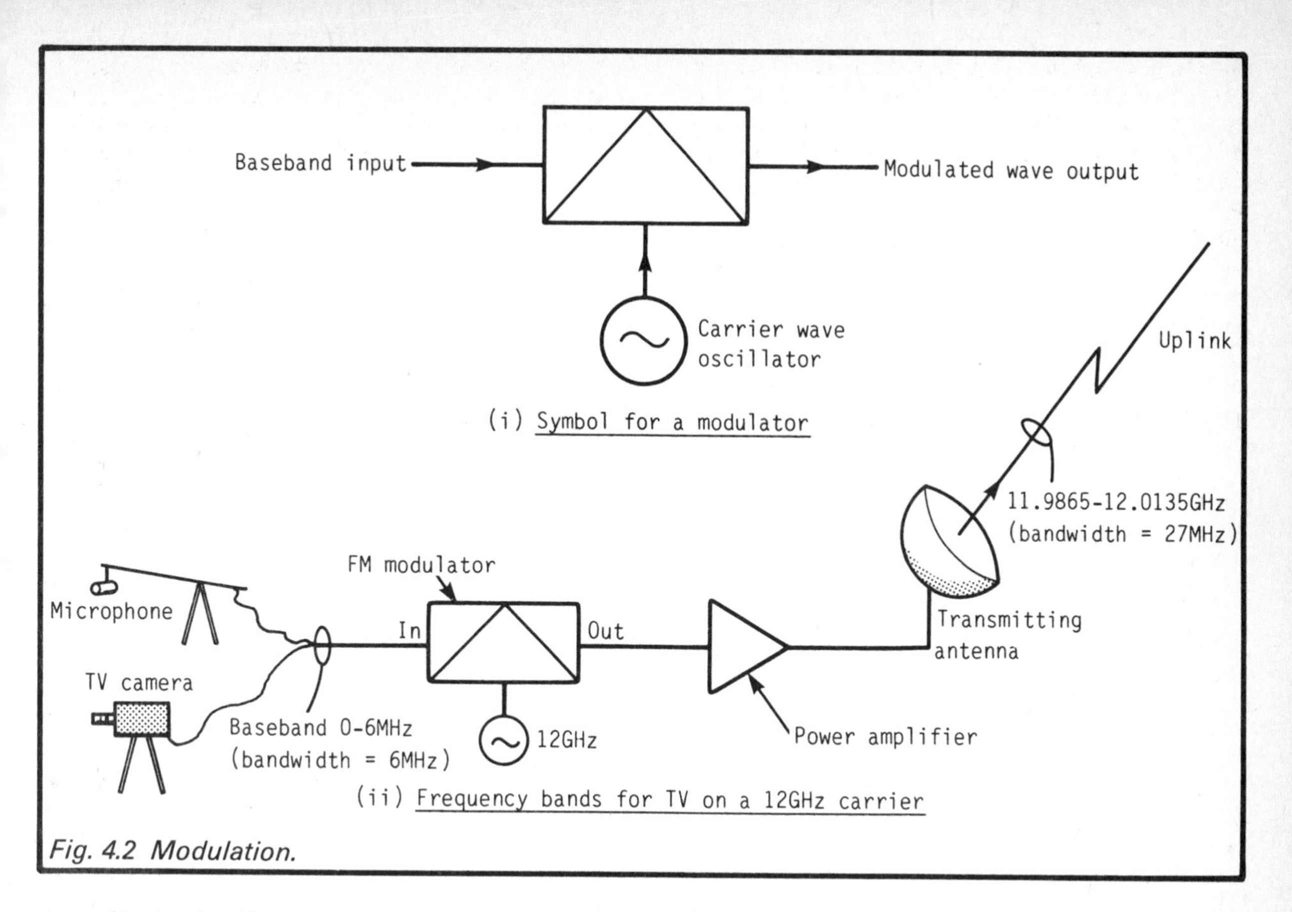

Fig. 4.2 Modulation.

waves (the baseband)

(ii) satellite tv uses frequency modulation (fm)

(iii) transmission of information of any kind requires a band of frequencies and a channel must be capable of transmitting this band fully for faithful reproduction.

4.4.1 Satellite Channel Bandwidth

It can now be appreciated that in Section 3.4 are quoted nominal or centre-channel frequencies only. Working things out for the full tv baseband signal (more on this in Chapter 7) which extends from zero up to 6 MHz, an fm wave in its need to accommodate sufficient of the harmonics, requires a bandwidth of about 27 MHz. This is the bandwidth recommended for European DBS. Every link in the chain must therefore be capable of transmitting at least this band, if one is not then the whole channel is restricted. Figure 4.2(ii) puts some typical figures showing the frequency bands as might be transmitted up to a satellite. The downlink right through to the demodulator in the satellite receiver equally accommodates a 27 MHz bandwidth.

4.5 Frequency Changing

Chapter 6 explains why a satellite up-link frequency differs from that of the down-link. That this can be accomplished indicates that some device is available

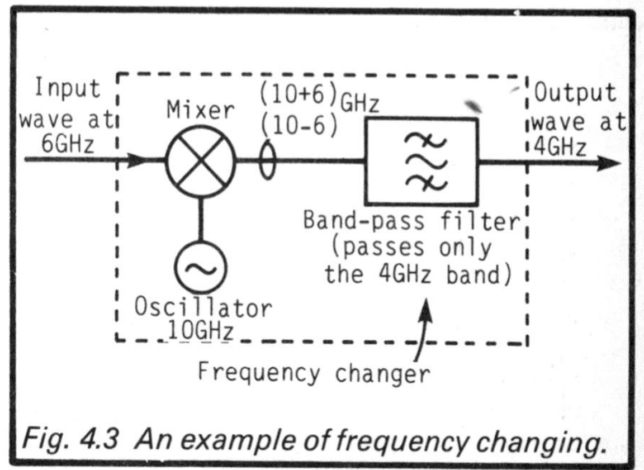

Fig. 4.3 An example of frequency changing.

for shifting the tv baseband signal from one carrier wave to another at a different frequency. It is known appropriately enough as a *frequency changer.* Such a device relies on the fact that if a lower frequency signal f_1 is mixed in a certain way with a higher frequency f_2, then two other frequencies are produced, $(f_2 + f_1)$ and $(f_2 - f_1)$, the sum and difference. One of the two new components is rejected by use of a *band-pass filter.* This is an electronic circuit within the frequency changer which passes a certain band of frequencies but no others. Suppose it passes $(f_2 - f_1)$ hence rejecting $(f_2 + f_1)$ and consider that within a satellite an up link frequency of 6 GHz has to be changed to a down link frequency of 4 GHz. Mixing the 6 GHz wave with the output of a 10 GHz oscillator then gives (10 – 6) = 4 GHz. Modulation impressed on the 6 GHz wave is carried through onto the 4 GHz so the same information is there but on a carrier wave of a different frequency. This is shown in more detail in Figure 4.3 which also illustrates the basic components of a frequency changer. The general symbol is given in Figure 4.1(iii).

Because higher frequencies are more difficult to handle than lower ones, frequency changing down is often employed. This is almost invariably the technique used in radio receivers (including the "tranny") and Chapter 9 shows that the satellite receiver is no exception. This is also the main function of an LNB, i.e. to convert or frequency change the incoming signal to a lower, more manageable one.

Chapter 5

SATELLITES – GETTING INTO ORBIT

Study of satellites naturally brings into focus Nature's way of holding the Universe together which was a great source of puzzlement throughout the early ages. Eventually in the mid-sixteen hundreds, the English physicist, Sir Isaac Newton, discovered that what kept planets in their appointed places and ensured that people were not thrown off the spinning Earth, was the force of gravity (Sect.2.4.2). From then on the fund of knowledge concerning gravity, forces and motion grew apace and it is from this that the creation of artificial satellites has become possible.

5.1 Going Round

A satellite has first to be got up there and then made to go round. It is perhaps better if we reverse the order and first decide how they stay in orbit, rather than how they reach it. Their travelling in far-flung celestial circles has many parallels with the motions of the planets round the Sun but not with comets, most of which have orbits which are anything but circular. Fortunately the tv satellite does have a circular orbit so enabling us to begin learning about the process in a more simplified fashion.

5.1.1 Motion in a Circle

We are all accustomed to going round in circles, both metaphorically speaking and in practice. The latter is of greater interest so here are just two examples. Firstly we all have experienced travelling round a bend in a car. If the car moves round left, the passengers lean to the right because their bodies have a distinct preference for continuing straight on. Again, if a stone tied to the end of a string is swung round in a circle, all is well until the string is released whereupon the stone flies off in a straight line.

From these examples it is evident that things only acquire circular motion if some outside force continually pushes or pulls them towards the centre. The force required is known as *centripetal* (centre-seeking) and in the case of the car it is provided by the grip of the tyres on the road so ensuring that the car moves round the bend. The force in opposition to this is the one acting not only on the passengers but also trying to make the car skid, known as *centrifugal* (centre-fleeing). With the sling, the centripetal force pulling the stone towards the centre of the circle is provided by the string, that there is tension in it is obvious because it is taut.

Newton wrote down Nature's rules for such motion. His *First Law* tells us that:

> "a body will continue in its state of rest or of uniform motion in a straight line unless acted upon by some external force".

Hence for a body to go round in a circle in preference to going straight on, somehow a force must be continually pushing it towards the centre.

From the general formula [A5(1)] we find that the force required is directly proportional to the square of the speed of the moving object and inversely proportional to the radius of the circle. This shows that as speed increases, the likelihood of a car skidding increases greatly and it does also the tighter the bend.

5.1.2 Motion in Orbit

Some medieval mechanics spent their whole lives trying to invent a *perpetual motion machine*, that is, one which runs without the input of fuel or outside aid. They failed because such a device would operate against the laws of Nature. Friction losses take energy away and turn it into heat, and the lost energy must be replaced. This is how things are on Earth for whatever moves on land, it has to do so through air. The minute gas particles flow over the body creating friction and energy has to be provided otherwise the body comes to a halt. Up in space, however, things are different, there is no air to cause friction so bodies, once given motion go on seemingly for ever, just as Newton predicted (Sect.5.1.1). Thus a bullet fired on Earth gradually loses speed until it eventually falls. Up in space, given a clear passage and freedom from gravity it will continue its flight in a straight line and never stop. But we demand that a satellite goes round in a circle so must now apply to it a centripetal force. Here at last is perpetual motion but not on Earth, only in space.

Gravity comes to our aid. To maintain the desired orbit, the centrifugal force created by the satellite's motion (which tries to pull it further away from Earth) can be exactly counterbalanced by the centripetal force provided by gravity. This latter force decreases as the distance from Earth increases so it is possible to adjust satellite speed and height for such equilibrium. Put in another way, the nearer to Earth a satellite is, the greater the speed it needs to avoid being pulled down. This indicates that satellites can operate at any height but for geostationary working there is an additional factor to take into account, the orbit time. The calculations to determine the height and velocity of a geostationary satellite are reasonably straightforward as shown in Appendix 5 (Sect. A5.2). Note that we work to 2 or 3 places of decimals only. There is no point in going further because the calculations are only designed to demonstrate the principle. In the practical situation very many corrections are made, even an allowance for the changing positions of other planets. Hence, approximately height of satellite above equator = 35,786 km

velocity of satellite = 11,069 km/hour .

5.2 Lift-Off

Nowadays a familiar but still awe-inspiring sight is the launch of a rocket. Tonnes of fuel are burnt every second[A5(5)] with the inevitable reaction of "will it or won't it?". But most rockets do. A rocket is propelled by the force generated by escaping hot gases.[A5(6)] These are produced by burning a combustible fuel with an oxidiser, the latter supplies oxygen for the burning process in the absence of air. There is a range of fuels in use, e.g. kerosene (paraffin), liquid hydrogen, ethyl alcohol for which liquid oxygen is frequently used as the oxidant but there are others such as liquid fluorine and nitric acid. The liquid gases are stored and loaded at very low temperatures.

The main requirement of a rocket motor is *thrust*, i.e. the upward push.[A5(4)] Thrust depends on two factors, (i) the exhaust gas velocity,[A5(6)] i.e. the speed at which the gases are impelled downwards out of the exhaust and (ii) the rate at which fuel is consumed.[A5(4)] Figure 5.1 shows a propulsion system in outline. The oxidiser and fuel are pumped into the combustion chamber and the mixture ignited. Hot gases expand and are forced out of the bottom of the exhaust nozzle giving rise to the billowing cloud of steam and smoke so evident in any rocket lift-off.

There are two types of rocket carrier systems:

(i) expendable launch vehicles (ELV). These are unmanned and the various stages of the rocket detach themselves when their work is done. These parts either burn up in the friction of the atmosphere or fall into the sea leaving the satellites to continue the journey. A rocket of interest to Europeans is the ARIANE, the pride and joy of ARIANESPACE (Europe) whose launching site is at Kourou, in French Guiana. An ARIANE can carry and launch two satellites, the payload is in excess of 4 tonnes;

(ii) space transportation system (STS). This carries astronauts who glide back to Earth when the mission is completed. The one known to all is the SHUTTLE ORBITER. It is as large as a modern jet-liner with a payload of nearly 30 tonnes. It is blasted into space by a huge rocket which is detached and discarded in stages. The SHUTTLE then continues its journey, steered as necessary by rocket motors. Satellites are carried in the cargo bay and are ejected by the crew into space whereupon ground control takes over. The SHUTTLE is operated by NASA[A7] from Cape Canaveral and can be re-used many times.

Sketches of ARIANE and SHUTTLE are given in Figure 5.2.

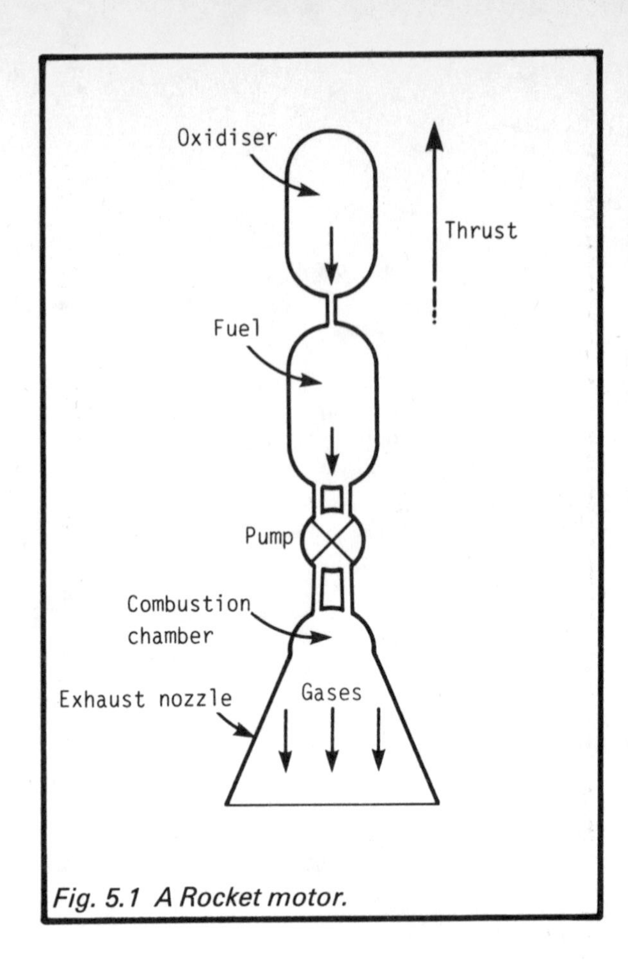

Fig. 5.1 A Rocket motor.

5.3 Into Orbit

Placing a satellite in orbit is not just a question of straight up and turn right. Launching is an expensive and extremely complex operation and practical considerations demand a technique involving at least one intermediate orbit.

Most launchers eject the satellite into an inclined *elliptical* transfer orbit first. Nature herself uses both circular and elliptical orbits for the planets and comets and so she has cleverly arranged the mathematics to suit. Thus we find that a circle is simply a special kind of ellipse.[A5(8)] Technically an ellipse is produced as in Figure 8.2(i) but in a simpler way it could be considered as a circle in a squeeze. Similarly with artificial satellites, most have elliptical orbits, the geostationary being an exception. We may be puzzled as to how the centripetal and gravitational forces balance (Sect.5.1.2) when the distance of the satellite from Earth is constantly changing. However, equation A5(9) shows that the velocity is not constant and in fact it varies in such a way that the two forces balance whatever the distance, e.g. the nearer the satellite is to Earth, the faster it goes.

Launching is preferably from a site near the equator because there the surface velocity of the Earth is greatest. In fact while still on the ground the rocket and its payload are already travelling in the required direction. The effect is comparatively small

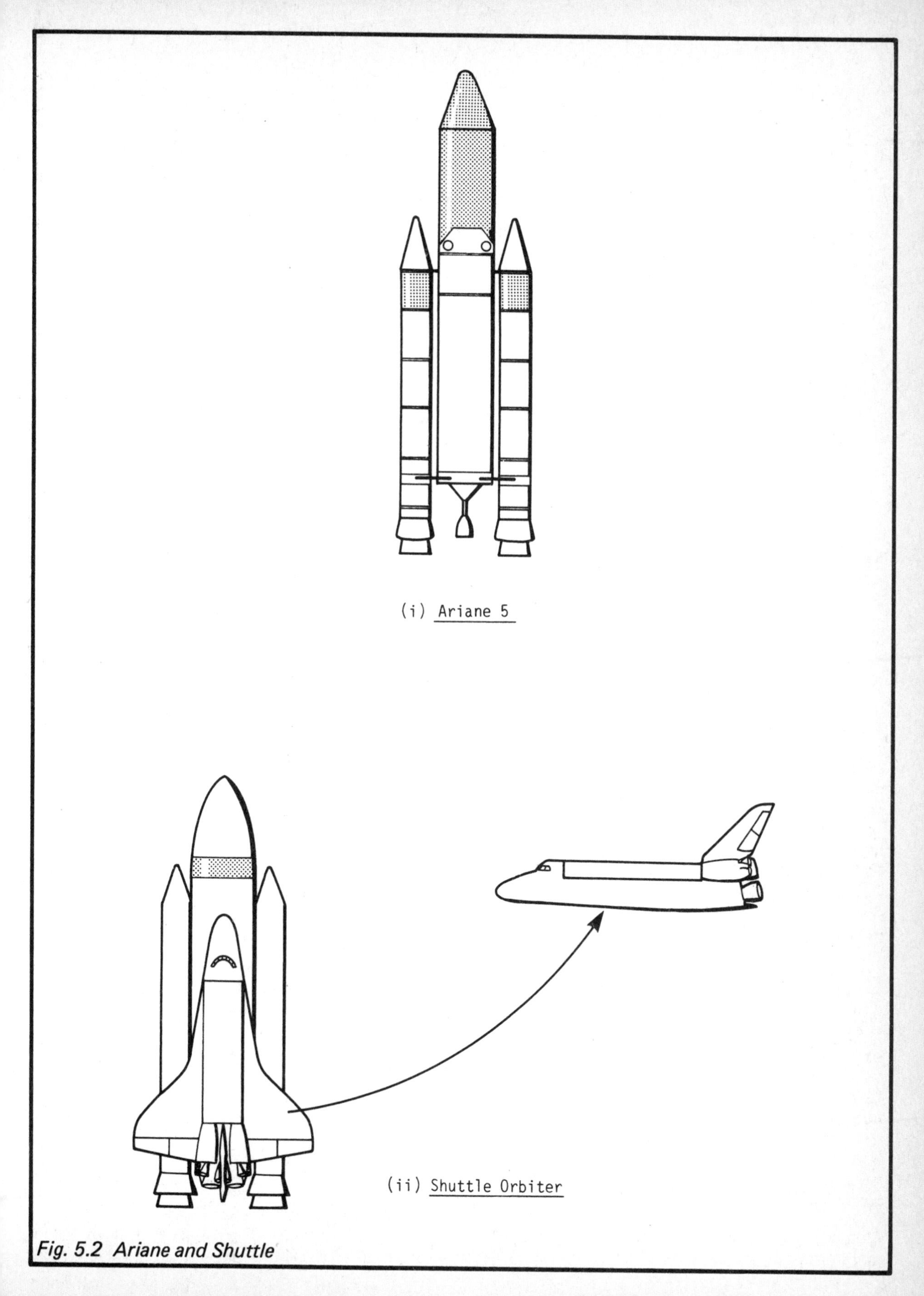

Fig. 5.2 Ariane and Shuttle

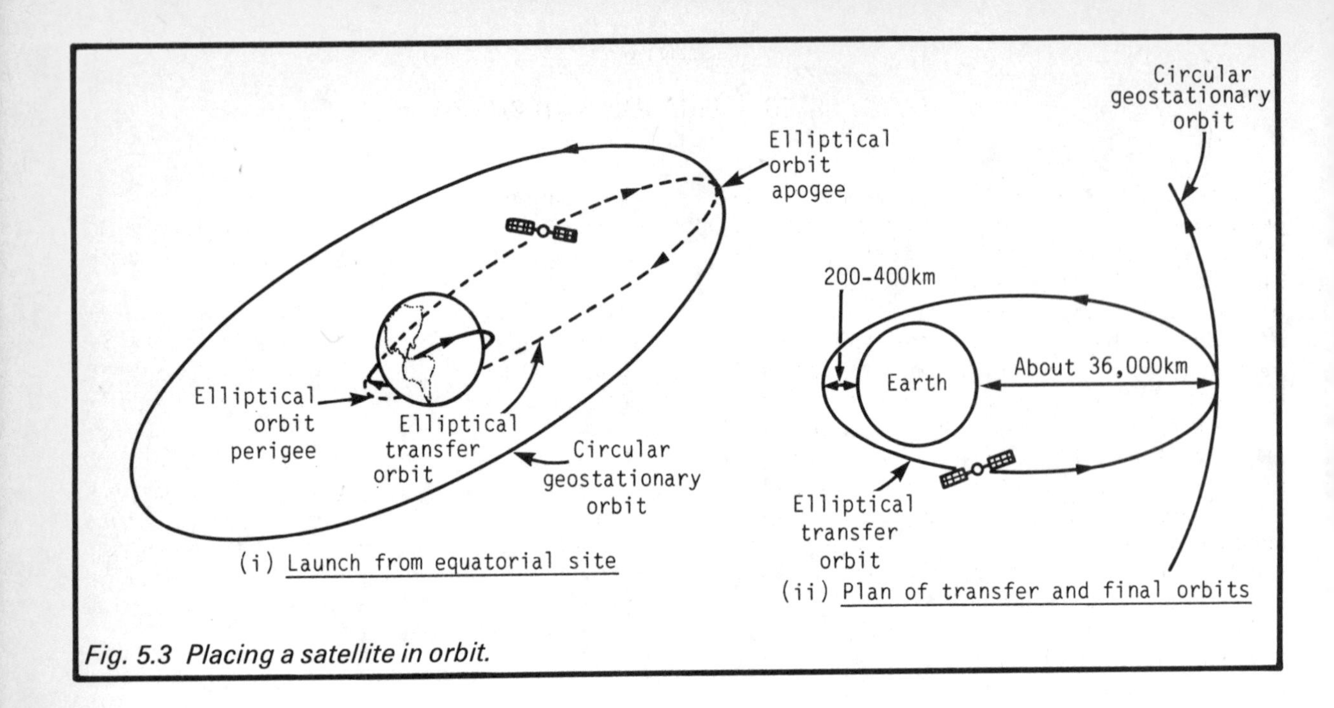

Fig. 5.3 Placing a satellite in orbit.

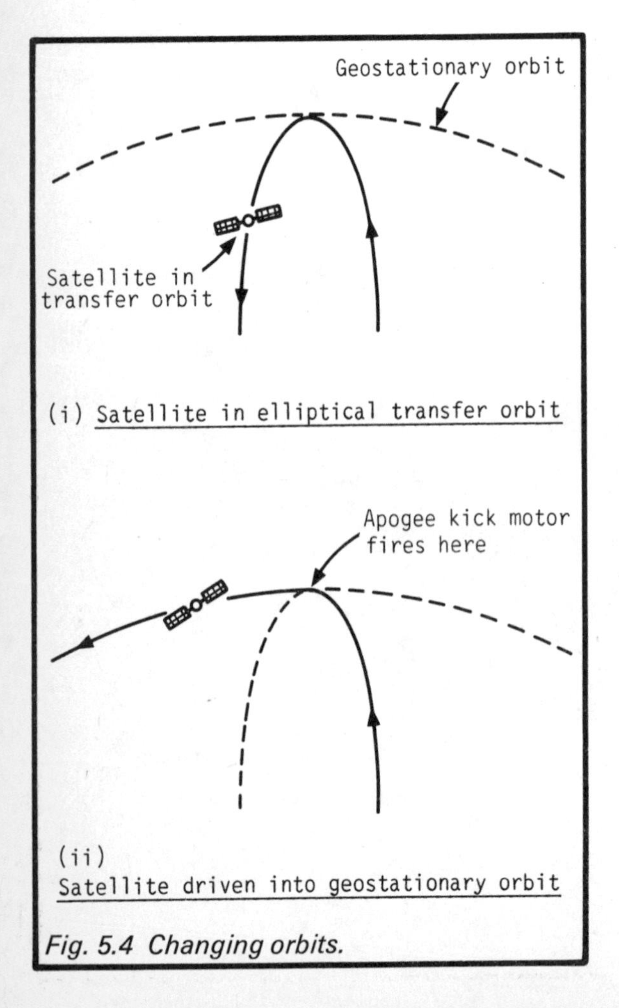

Fig. 5.4 Changing orbits.

but worth having. At the equator we travel at over 1600 km/hour (over 1000 miles/hour) without even knowing it![A5(10)] Kourou is an example, it is only 5° of latitude away from the equator.

The transfer orbit which ELV launchers use has its perigee (the point on the orbit nearest the Earth – see Fig.5.3) quite close at some 200–400 km high. The apogee (point furthest away) is at the geostationary altitude. The inclination of the transfer orbit may vary with the latitude of the launching pad but it cannot be less than the latitude. The satellite stays in the transfer orbit as required until, when at the apogee, a rocket motor on it is fired (the *apogee kick motor*, AKM). This moves the satellite out of the transfer orbit into the geostationary as shown in Figure 5.4.

In a launch by the SHUTTLE, the transfer orbit is not reached directly, instead the SHUTTLE moves first into a circular *parking orbit* some 100 km up. When at the perigee of the proposed transfer orbit the satellite is launched and a perigee motor on it is used to provide the boost for movement into the new orbit.

Chapter 6

SATELLITES – THE BITS AND PIECES

At least satellite engineers are not harassed by the problems of air flow and its resistance as experienced by car designers. Accordingly satellites can be expected to come in all shapes and sizes with no streamlining. Widely differing they may be but there is usually some sort of central box with wings of solar panels pushing out to catch sunlight for turning into electrical power. Figure 6.1 shows the outline of a typical communications satellite. This one has a central rectangular box although cylindrical ones are also in use. During flight from Earth to orbit the solar panels are folded in to minimize the space required within the rocket. Typically a satellite might weigh (on Earth) 1800 kg (nearly 2 tons) and have a total span of about 20 m. The cost of building one and getting it into orbit runs into tens of millions of dollars or pounds. In addition there are the ongoing costs of looking after the device once it is up and flying, a most expensive affair altogether.

6.1 Control

Nothing is possible unless the satellite *attitude* (its position relative to what it ought to be) can be controlled precisely from the ground. Considering that a shift of a mere one degree in the aim of the transmitting antenna would move the footprint nearly 700 km (about 400 miles), it is obvious that stability is essential. Moreover a satellite must be kept exactly in its rightful position in the orbit despite the many outside influences such as gravitation of Sun, Moon and other planets which tend to move it away. In fact the Regulations require a station-keeping accuracy of ± 0.1° in longitude although with the realization that this represents a distance of 74 km (46 miles) on the orbit it perhaps does not seem so onerous.

There are two attitude control systems:

(i) *spinning* – the body of the satellite is

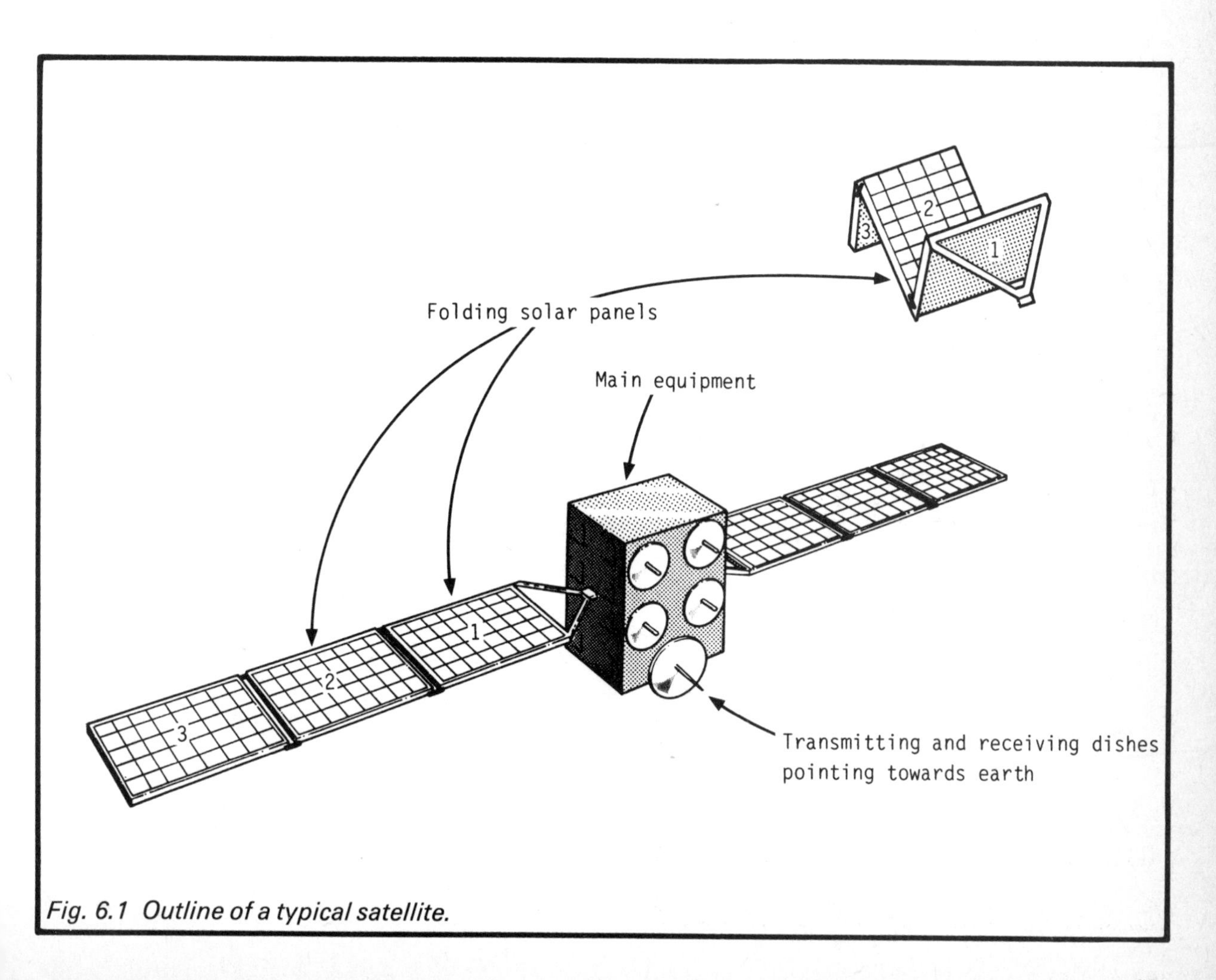

Fig. 6.1 Outline of a typical satellite.

cylindrical and rotates at up to 100 rpm and this gyroscopic action opposes any change in the spin axis. The same principle is in use in a ship's stabilizer in which a gyroscope checks rolling in heavy seas. In this system the antennas have to be *despun*, i.e. rotated in the opposite direction so that they always point to the same spot on Earth:

(ii) *axis stabilization:* three solid disc *momentum wheels* are rotated at high speed by motors within the satellite in the three mutually perpendicular directions. If any wheel is speeded up the satellite tends to turn in the opposite direction and vice versa. By controlling the appropriate motor speeds the satellite can be turned into any position.

Satellites are therefore monitored 24 hours per day by the ground control centre and signals are sent up to operate electric or rocket motors or gas jets so that attitude, station-keeping and antenna pointing are all correctly adjusted. To tell the Earth station exactly what is happening up there is a *telemetry* system (from Greek, measuring from afar) operating within the satellite, continuously transmitting data from the various sensors. These for example, measure the electrical conditions, especially the state of the power supply, amount of fuel left, fuel tank pressures, temperatures and sighting devices for determination of attitude. For tracking, control sensors measure velocity and acceleration. Modern equipment is capable of such sensitivity that a satellite can be pin-pointed to within about 100 m. Considering the distances involved, this is precision indeed.

For the transmission of telemetry data and reception of command (control) signals a separate satellite antenna is normally employed.

6.2 Transmissions

This is what a satellit exists for although in fact the equipment for doing the job many represent only a small part of the weight and cost of the complete unit. The tv programmes are sent up by the ground station on microwave carriers (Sect.4.4). This up-frequency ends its journey on a satellite receiving antenna, it is then converted to the satellite transmitting frequency for sending to Earth via the down-link. For European DBS the frequencies are within the range 11.7 – 12.5 GHz, each carrier having a bandwidth of 27 MHz. Change of frequency from up-link to down-link is essential for as shown in Figure 6.2(i), without a frequency change and highly directional though the antennas may be, some leakage of the powerful down-link signal will be picked up by the receiving antenna. This passes through the amplifier and emerges at enhanced level so increasing the level of the leakage. The effect is cumulative and the whole system becomes unstable. This happens occasionally with public address systems where the instability is heard as a "howl". A frequency change within the satellite avoids this because the receiving system does not accept the transmitted frequency.

To change frequency a local oscillator is required (Sect.4.5) and so, at its very basic, the equipment for one channel (one tv programme) can be illustrated as in Figure 6.2(ii) for the European DBS range. This satellite equipment, excluding the antennas is known as a *transponder* (transmit + respond -er). Several are carried by each satellite.

6.2.1 The Power Amplifier

Of the many strange devices used in satellite systems, certainly the travelling-wave tube (TWT or TWTA where the A stands for "amplifier") is one. These letters frequently appear in technical fact sheets of satellite transponders. TWT's are used for the power amplifier stage shown in Figure 6.2(ii).

Signal amplification at the very high satellite frequencies is far from easy when compared with lower frequencies as in the broadcast radio range. As frequencies get higher they become more difficult to handle and the normal circuit wiring techniques are no longer usable. A breakthrough came in the early 1940's when the TWT arrived and development of the device has continued ever since for greater output power, greater bandwidth and for use at higher frequencies. Basically it is coupled in between two waveguides (Sect.3.5), the output delivering much higher power than is applied to the input (as with any power amplifier). As shown in Figure 6.2(ii) the output waveguide takes the wave directly to the transmit antenna. In the TWT the wave passes along a kind of waveguide helix within an evacuated glass envelope some 60 cm long. Shooting down the centre of the helix in the same direction as the wave is moving is a narrow stream of electrons. The whole system is so arranged that there is mutual interaction between the wave and the electron stream and as the former progresses around the turns of the helix, energy is transferred from the electron stream to it. By this rather complex interaction the wave is amplified. This could be likened to sliding down a helter-skelter where gravity (the electron stream) continually adds to motion so that a relatively slow start at the top becomes a speedy exit at the bottom. TWT's can generate power outputs from a few watts to around 200–250 with a gain in excess of 50 dB.

Solid state devices are also usable. These consist of solid material as opposed to employing an electron stream shooting through a vacuum. They are really highly specialized transistors and although at present they have difficulty in competing with TWT's, their time will come.

6.3 Power

Power to drive everything electrical within a satellite is obtained from *solar cells.* Such a cell is manufactured from a thin layer of silicon or gallium arsenside on a special base. To obtain a large surface area for catching sunlight, the cells are usually spread

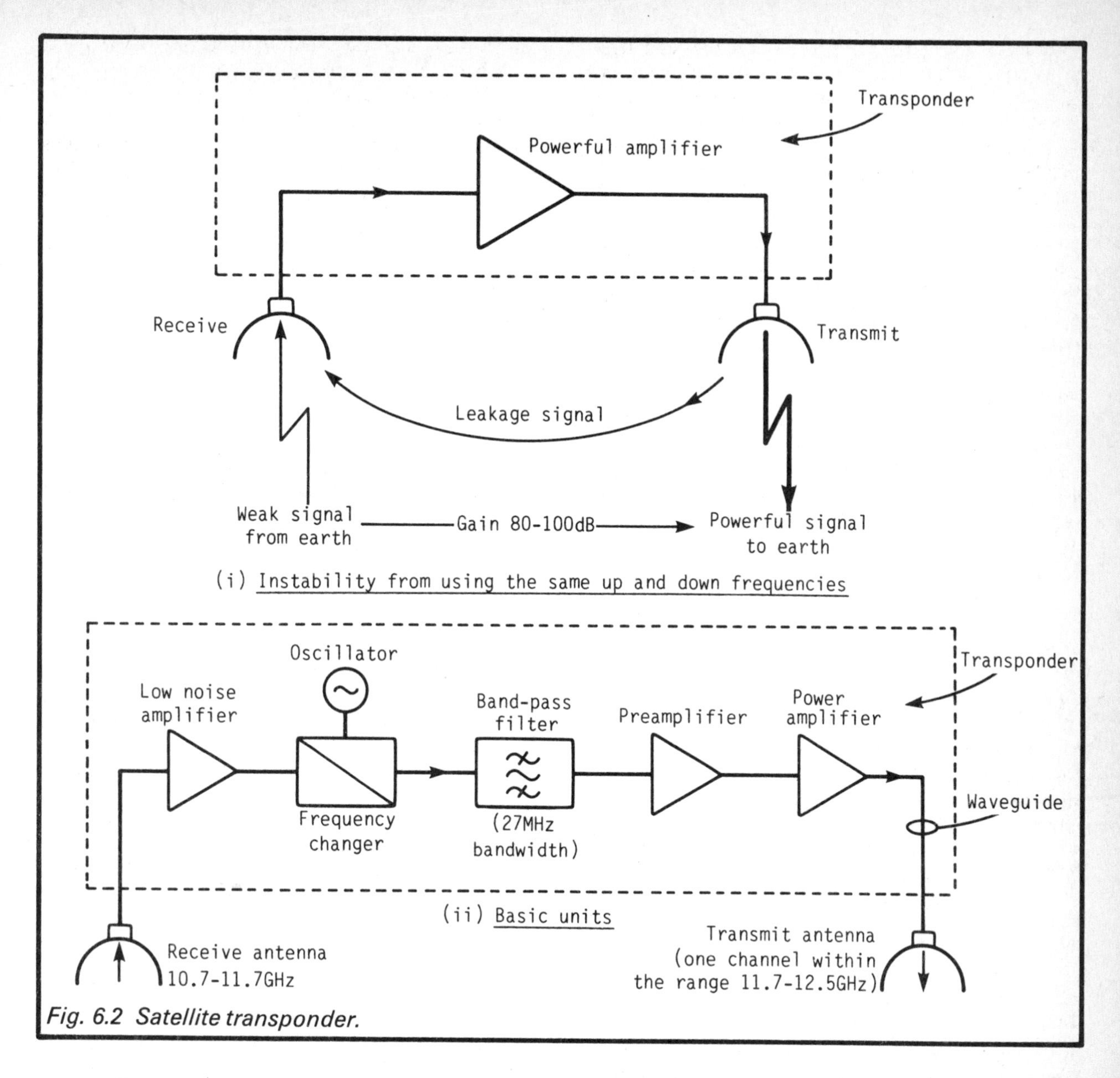

Fig. 6.2 Satellite transponder.

out on flat, wing-like panels, a distinctive feature of most satellites. Alternatively drum-shaped spinning satellites can conveniently have the cells fixed round the outer surface, no panels are therefore required. The electricity from a single cell is small, hence many cells are required.

Up in space with no filtering by the atmosphere, the Sun's rays are at their most powerful and in fact there is over one kilowatt of energy for each square metre (sufficient to light 10 x 100 watt lamps). Solar cells are relatively inefficient in converting this energy into electricity but they are improving with new developments. Early cells had efficiencies of only a few per cent but the latest exceed 20%. Light from the Sun therefore, in falling on the solar cells generates sufficient electricity to keep the satellite batteries charged. Those most used are of the nickel-cadmium type, the same as used at home to save continually buying batteries. They must never run down so the number of solar cells has to be ample. Furthermore satellites suffer an *eclipse* when in the shadow of the Earth as illustrated in Figure 6.3. There are two eclipses to contend with annually, both around the equinoxes (about March 21 and September 22 when day and night are equal). The eclipses start at 1 or 2 minutes per day some 20 days before the equinox, rising to more than one hour per day at the equinox then falling throughout the next 20 days. During an eclipse the satellite relies entirely on its batteries, hence the requirement of sufficient capacity. Modern satellites need a total power of some 1–2 kW. This may seem small when such an electrical power on Earth heats only one room, but it is a feat of modern engineering that all this power is extracted free from the Sun's rays.

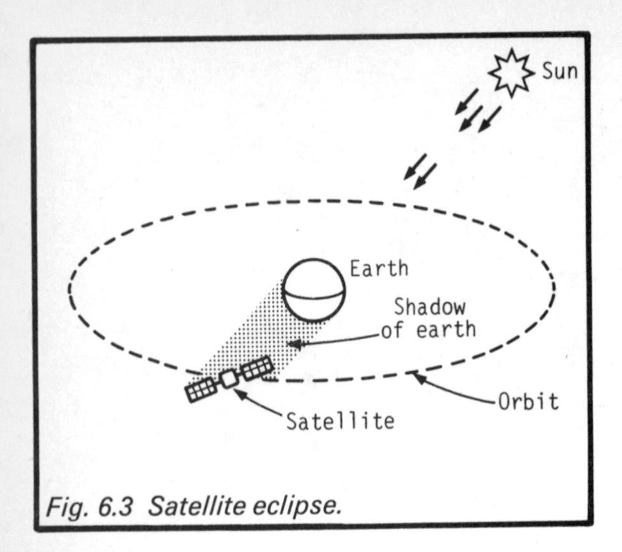

Fig. 6.3 Satellite eclipse.

6.4 Antennas

Satellites nowadays carry several antennas falling mainly into two categories, parabolic dish and horn. The former is covered in Section 8.1 which although concerned with receiving antennas, in fact describes the basic principles and shape of the satellite variety also. Horns of various shapes also have their part to play, sometimes for their omnidirectional properties. This is in contrast to the dish with its property of focussing the radio wave into a narrow beam. Three of the horn shapes which are likely to be used are given in Figure 6.4. These are attached to the ends of waveguides of appropriate shapes as shown.

It may seem a little odd, mentioning antennas with omnidirectional properties. Everything we have studied so far seems to require a narrow radio beam aimed at a particular satellite on an up-link or at a small area on Earth on a down-link. However while a satellite is being dovetailed into the final orbit its correct attitude with respect to Earth has not been established so the directional antennas do not point correctly. An omnidirectional antenna is therefore the only means of maintaining contact with the ground control (Sect.6.1) and it is used until the attitude has been stabilized. After this a directional antenna takes over.

6.4.1 Footprints

Imagine a powerful searchlight vertically above at geostationary height, then on a smooth earth (no mountains etc.) the pool of light is circular even though the surface of the Earth is curved. But suppose the Earth to be flat and that the searchlight is moved to one side so that its beam arrives at an angle less than 90°, the pool of light then becomes elliptical. This can be demonstrated by going on to Figure 8.2(i) where the top part of the middle cone represents the searchlight beam. This is all very well but the Earth is not flat and the beam can arrive at different angles with regard to both latitude and longitude. It can be shown (with a pocket torch and football if necessary) that under these practical circumstances the elliptical pool of light changes to oval (egg-shaped) with the tapered end on the polar side. The same applies to a satellite radio beam for which the area "illuminated" is the footprint.

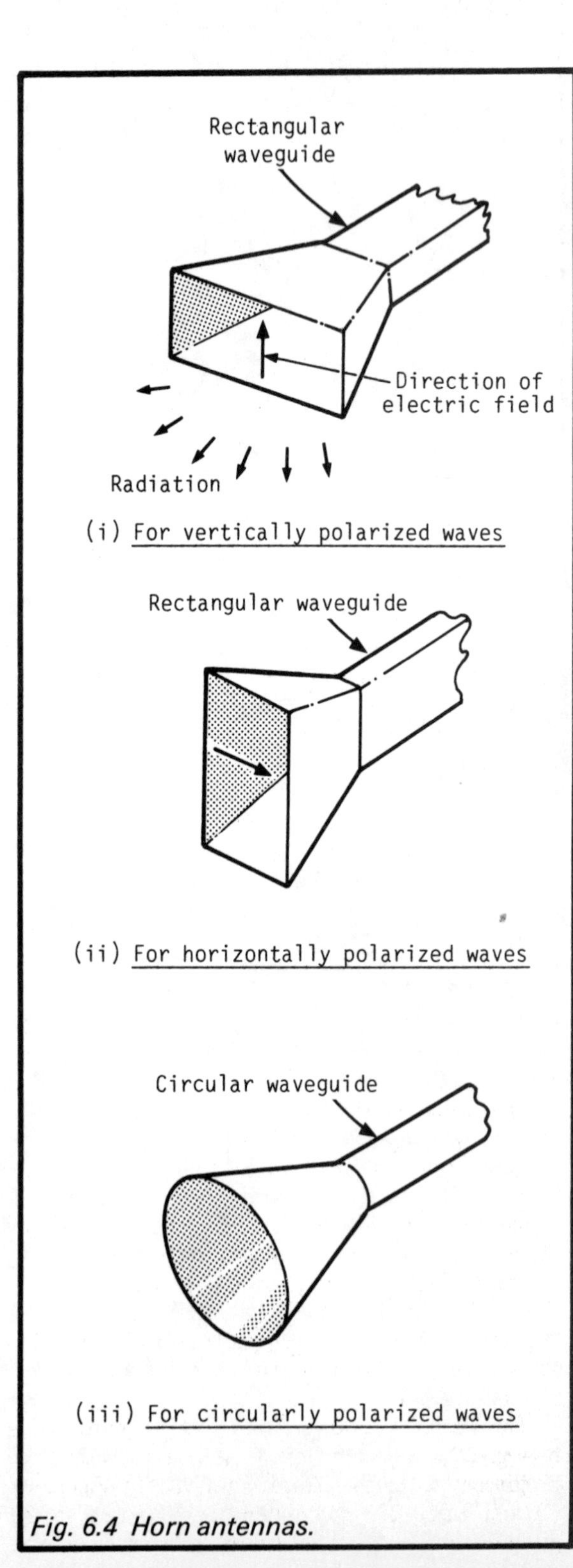

Fig. 6.4 Horn antennas.

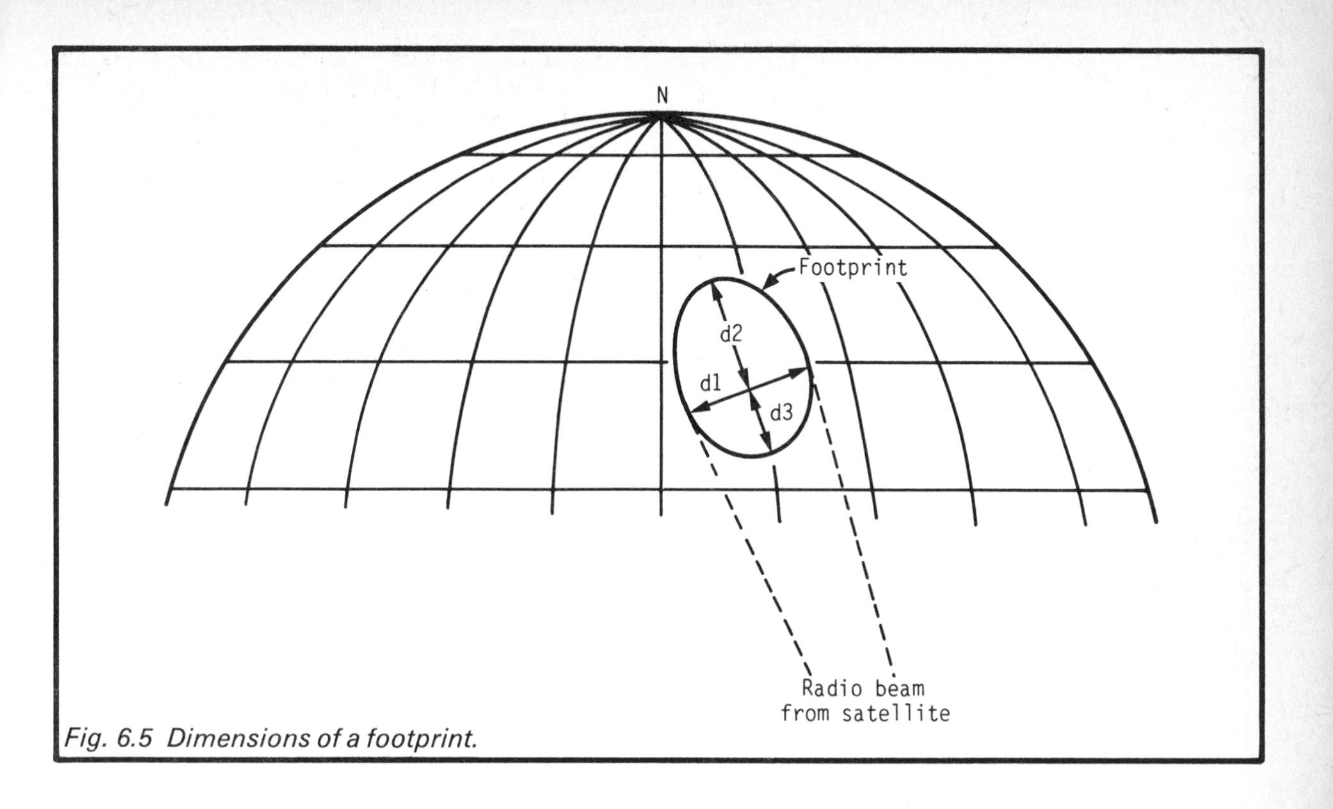

Fig. 6.5 Dimensions of a footprint.

Placing a footprint of the right size, in the right place and with the best radio signal strength is the work of one antenna or the combined efforts of several. Briefly they have the capability of focussing the radio wave into a slightly divergent beam and directing it down to Earth.

Through somewhat complicated trigonometry, certainly not recommended for those without computer or calculator, the theoretical footprint area can be calculated. Figure 1.2 illustrates how the footprint is formed and Figure 6.5 shows how three dimensions only are needed to define the area approximately, d_1 across the widest part of the oval, d_2 the longer axial length from tip to the line of d_1 and finally d_3 from this point to the base. Each d can be calculated for any position on Earth provided that the "beamwidth" of the satellite antenna is known. This however is not likely to be quoted in general satellite performance literature because footprint maps are more useful. Typically they are as shown in Figure 6.6, the oval shape is evident. Frequently it is modified by the addition of spot beams (Sect.6.4.2). A figure such as this tells a user approximately what the signal strength from a particular satellite should be within the service area shown. Signal strengths are discussed in Section 8.3.3.

The size of a footprint is of course controlled by how greatly the beam from a satellite spreads out on its journey to Earth. This is illustrated by Figure 6.7 (i) and (ii). The maximum signal power on the ground is in the footprint centre, the power decreasing with distance away. Footprints are published in various ways but most frequently as an area delineated by a single line (as in Fig.6.6). This line represents the contour around which the signal power is half what it is in the centre, i.e. where it is "3dB down". A more complete footprint map includes further contours for lower levels of signal. For the "3dB down" contour the beamwidth angle is known as the *Half Power Beamwidth* (HPBW) as shown pictorially in Figure 6.7(iii).

So far so good, but what is also important here is that, for a given frequency, the beamwidth angle is directly controlled by the size of the parabolic antenna. Large antennas have narrow beamwidths and vice versa. This is one reason why satellite antennas are somewhat larger than one would expect considering that they have to be transported up there. Our calculations[A6(1)] would show that at 12 GHz, a 3 m dish would theoretically have a half power beamwidth of 0.5° resulting in a service area some 300–400 km across. In practice some allowance is necessary to allow for the fact that satellite "station keeping" is not perfect.

6.4.2 Spot Beams

Figure 6.6 shows primary footprints from single antennas. Many satellites however employ additional antennas to produce *spot beams*. These may be required to complete the coverage of a region of awkward shape for clearly not all fit into a normal oval. There also may be small areas of particular importance which must not be left out or for which

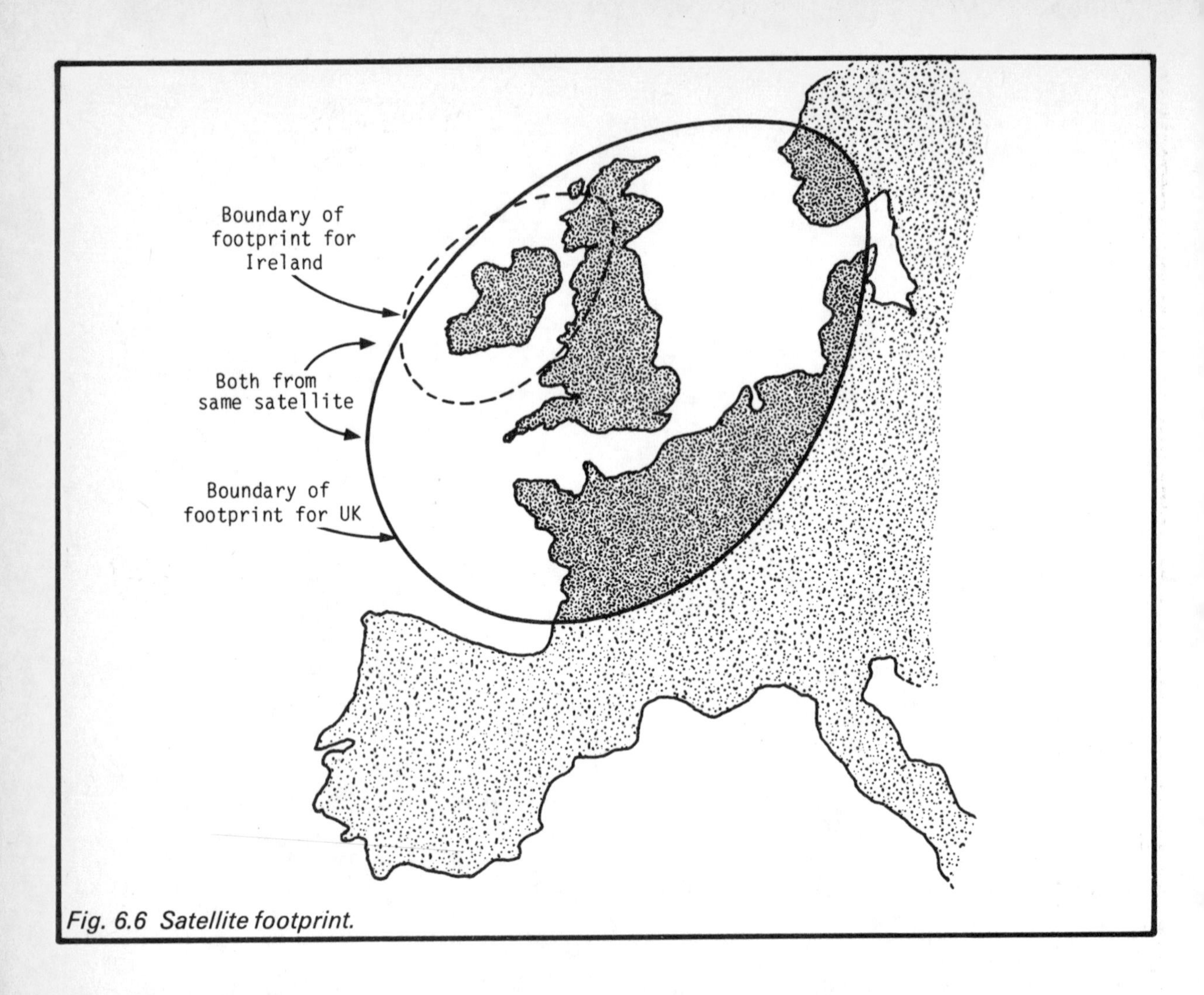

Fig. 6.6 Satellite footprint.

the received signal would otherwise be too weak. By using larger dish antennas than usual the radio signal can be concentrated into a narrower beam (Sect. 6.4.1). This provides a small added footprint which can be aimed at the region or spot required. Several such spot beams may work in conjunction with the main one. They are recognized by names such as "West Spot Beam", "East Spot Beam" etc. and their footprints may be completely separate from the main one or used to modify it.

6.5 Lifetime

Considering the cost of a satellite, naturally the aim of the design team is that it should live to a ripe old age. Some unavoidable factors combine to prevent this:

(i) present day solar cells (Sect.6.3) deteriorate and their efficiency falls with time;

(ii) the fuel supply for attitude correction and other manoeuvres eventually becomes exhausted;

(iii) there is always the risk of component failure, although this is very small indeed, especially as there is no atmospheric corrosion as there is on Earth. It is made even less catastrophic by duplication of systems so that if one fails the other automatically takes over;

(iv) there is one other potential source of disaster which is worth mentioning although so far it has given practically no cause for alarm. This is from meteorites and man-made debris in the shape of discarded items. Small meteorites usually burn up in the atmosphere so cause little trouble on Earth but they can be a hazard at geostationary heights. Man-made debris is on the increase, items may settle into orbits of their own so when such an orbit intersects the geostationary one there could be a collision. No problems so far but the risk is there.

Generally at least a ten-year life span is expected.

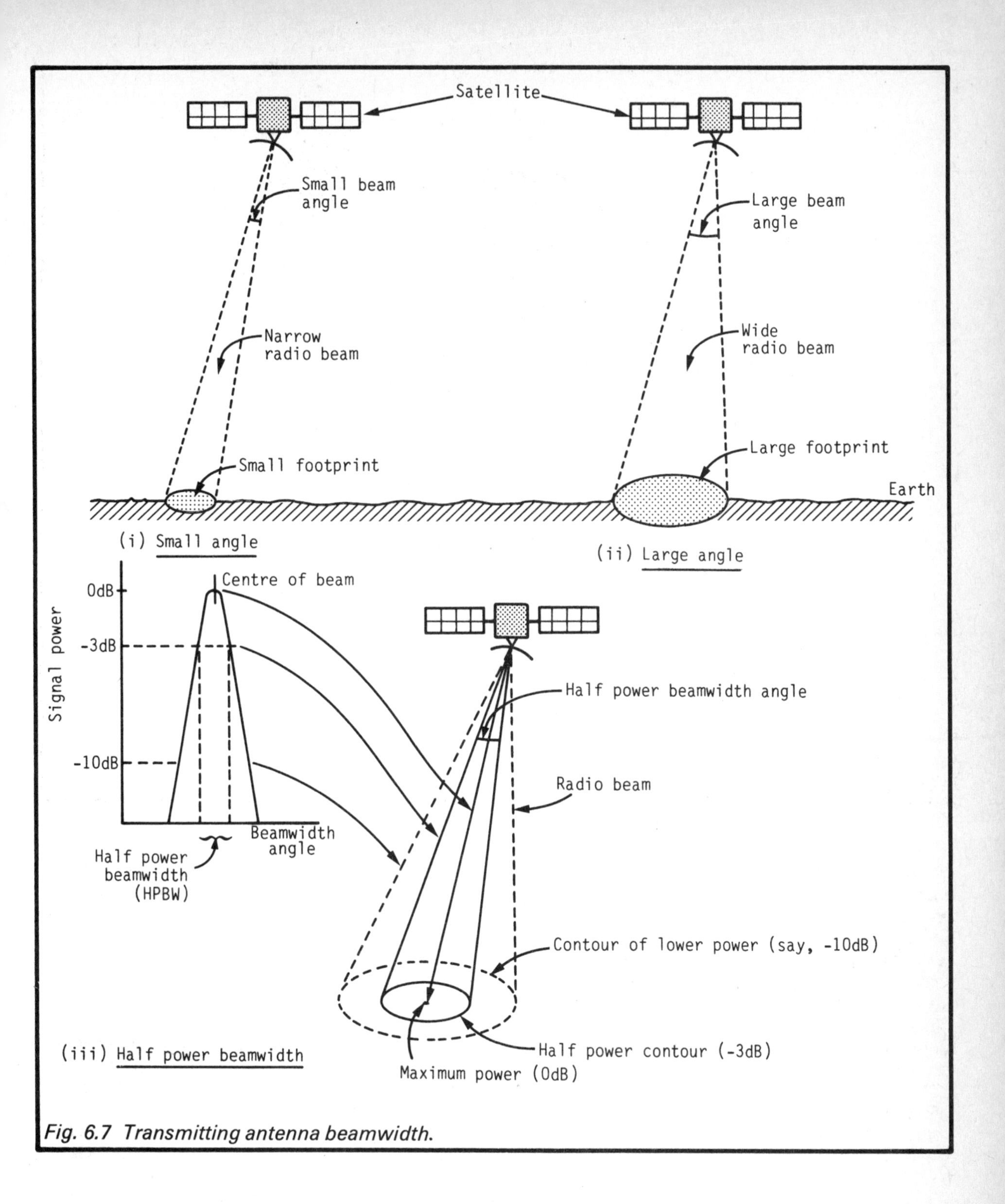

Fig. 6.7 Transmitting antenna beamwidth.

Chapter 7

TELEVISION

This resident thief of time had its early beginnings in the mid-1920's. However tv really took off when colour came along in the 1950's whereupon a technology already not that easy to understand by the layman became even more difficult. Nevertheless even though the system overall is complicated we must find out a little about it to understand the changes which are coming and also the significance of the mysterious MAC's which seem to have no relationship with our Scottish friends.

TV is the "package" which satellite systems carry and for any delivery system there is a limit to the number of packages which can be loaded, depending on their size. The package is of course, bandwidth (Sect.4.3).

7.1 The Flying Spot

Television pictures are built up by the efforts of a single tiny spot of light. Some idea of its actual size may be judged from the fact that if about 600 of them were stood on top of each other in a vertical line, this would just reach from top to bottom of the screen. Figure 7.1 shows in broad outline what happens within a television cathode ray tube (CRT). At the end remote from the screen electrons are ejected from an *electron gun* towards the screen. The latter is at a high positive potential (several thousand volts), creating a field which attracts them with a force sufficiently great that they accelerate to a very high speed, in fact to some 10^8 m/s (an unbelievable 200 million miles per hour).

The faster an object moves, the more energy it has and this energy must be given up if the body is stopped. The energy is known as *kinetic* (of motion). We see the proof of this when two cars collide at a walking pace, they do less damage to each other than they would at full speed. Kinetic energy also depends on the mass of the object, the formula is quite simple:

$$\text{kinetic energy} = \tfrac{1}{2}\, m.v^2$$

where m is the mass of the object and v is its velocity.

Electrons therefore, although of insignificant mass, reach such high speeds that they gain greatly in energy. Accordingly when many millions of them reach the screen together over a small area, each giving up its tiny packet of energy, the total is sufficient to cause the phosphor to glow brightly. We never see the spot because it is always kept on the move. In use the spot position on the screen can be anywhere according to the whims of the *deflection system*. There is more to a CRT than this but we need not go into detail. At this stage it is worth

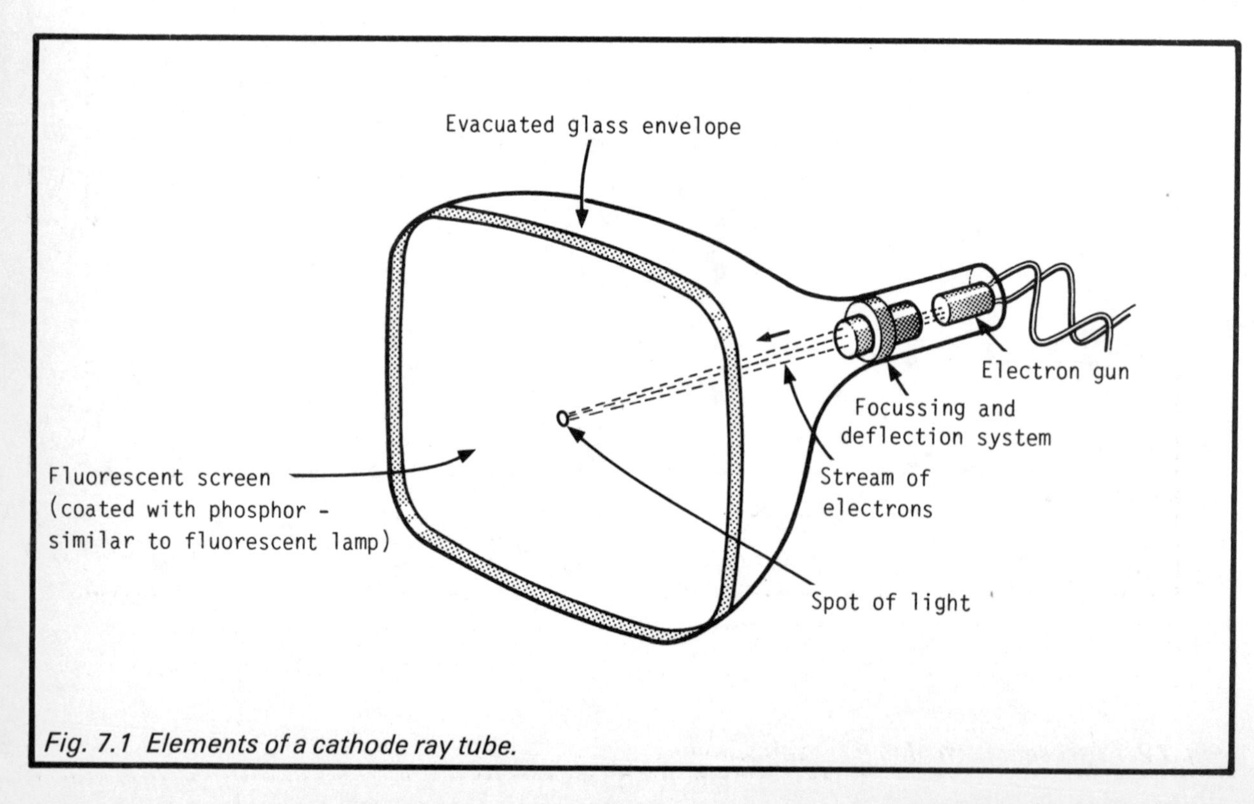

Fig. 7.1 Elements of a cathode ray tube.

noting that the spot can be light or dark according to the number of electrons in the stream.

7.2 Scanning

Consider what is on this page and how we take it in. The eyes *scan* the print, line by line, left to right. On completion of each line they move swiftly back to the beginning of the next, in tv terms this is known as *flyback*. The horizontal scanning across the lines is combined with a much more leisurely movement down the page. A page in a book is equivalent to a *frame* or *field* in tv (the term "frame" is generally used here to avoid confusion with the "fields" of Section 2.4.3, although in doing so we may not always be technically accurate).

Basically this is how a tv camera *sequentially* scans a scene. It examines each tiny area or *picture element* in the same order, left to right at the top, then after a rapid flyback, left to right on the line below for 625 lines (525 in some countries) to complete a frame of one still picture. This could be repeated 25 times each second so that when the still pictures are reproduced at the receiving end, the viewer has the illusion of movement. This is because what the eye sees lingers on for a short time (persistence of vision) so each full picture appears to merge with the previous one. It is the principle on which the cinema works and really we do not go to the "movies" but to the "stills". Unfortunately on the smaller tv screen, 25 pictures per second results in a noticeable *flicker* for most viewers. In contrast, experiments have shown that 50 pictures per second provides motion without flicker.

Normally doubling the number of pictures per second would double the information rate and therefore the bandwidth (Sect.4.3). This would be a high price to pay simply to reduce flicker so the technique of *interlaced scanning* is used. Instead of transmitting each line in sequence, the odd lines are scanned for one frame and the even for the next, giving two half-scans per picture. The effective presentation therefore is at 50 images per second with flicker reduced accordingly.

Considering next the home tv screen, if the tiny spot of light is controlled in this way, the result is a bright white rectangle the size of the screen. A representation of the technique using interlaced scanning is shown in Figure 7.2. Only a few lines are in each diagram and it is left to the imagination that the bottom of the spot on one line coincides with the top of the spot when traversing the line below. Such a pattern of scanning lines is known as a

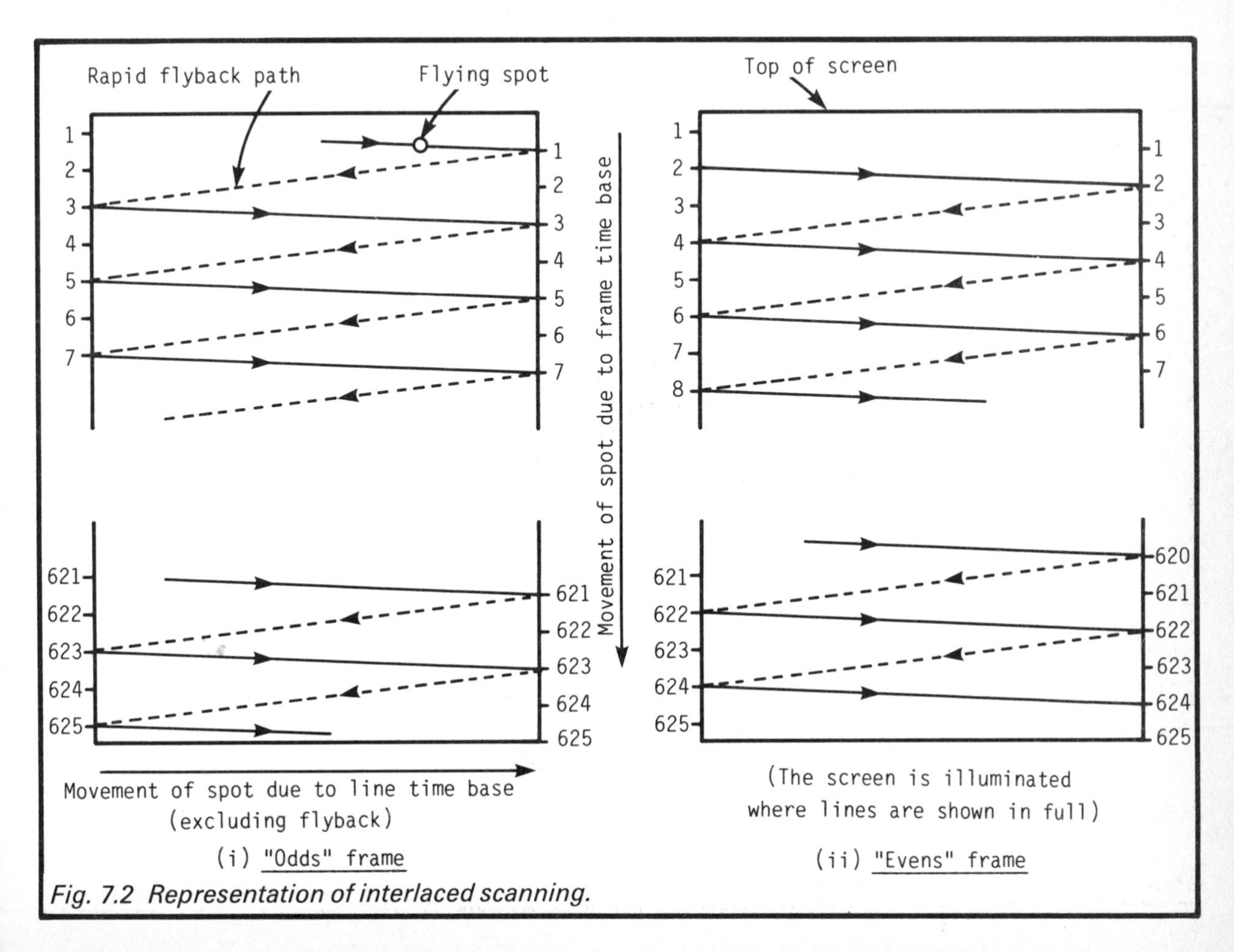

Fig. 7.2 Representation of interlaced scanning.

raster. An electronic *line time base* moves the spot across the screen and then lets it fly back quickly. At the same time a *frame time base* is moving the spot down. This is why on Figure 7.2 the lines slope downwards. As each frame is completed, the spot returns to the top and while this is happening, because no picture information is being received, the transmitter seizes the opportunity to send its *teletext* data. Figure 7.2 is drawn to show the idea of interlaced scanning, it is not an exact representation especially in that the number of lines on a working tv screen is a few less than 625, a little of the time is required for other purposes.

Of interest later for Figure 7.4 is the time taken to complete one line in a 625-line system. This number of lines is run through in $\frac{1}{25}$ seconds, hence:

$$\text{time for one line} = \frac{1}{625 \times 25} \text{ seconds} = 64\ \mu s$$

(μs = microseconds).

7.3 Synchronization

If a tv camera is scanning a scene and tv receivers are reproducing it, their scanning arrangements must work in exact unison, i.e. when the camera scanner commences a frame, the receivers must also. Cameras and receivers must be in *synchronism*. The system does even more, not only do the frames start in synchronism but so does every line. This is achieved by sending a short duration *triggering pulse* which is a tiny burst of energy causing the line time base to start drawing the spot across the screen. The pulse is only about 5 μs long occurring as shown above every 64 μs. Hence when the camera commences scanning a particular line, all tv sets start to reproduce the same line. When the line is completed the spots fly back to the left (Fig.7.2) to await the next triggering pulse to start the line below – rather like anticipating the starting pistol for a race. Similar arrangements shift all spots from the bottom of the screen to the top to await triggering for the start of the next frame.

7.4 The Picture

Next, a device which adjusts the number of electrons in the beam according to the brightness or *luminance* (from Latin, light) of each picture element is all that is required for black and white (monochrome) tv. With no electrons there is no spot, i.e. black; with the full quota (adjusted by the *brightness* control of the set) there is white. In between are all the shades of grey. Provision of this facility is quite simple. Within the CRT a *grid* surrounds the electron beam at the gun. When the camera sees white the signal it sends out eventually arrives on the grid and allows the full electron beam to pass. Anything less than white puts a negative potential (Sect.2.8.1) on the grid, repelling electron flow and producing a less than full-white spot.

We can now look at a graph of the video signal for one line of monochrome tv. This is not the radio signal but what it leaves behind on demodulation (Sect.4.4). Figure 7.3(ii) shows the output from the camera which is also the input to the CRT's of the receivers. For ease of understanding, the scene chosen is a clear chessboard mounted on a black background [Fig.7.3(i)]. The synchronizing pulse arrives first and is directed to the line time base which it starts. The remaining waveform passes to the CRT grid to control the brightness of the spot. Because the board is in black and white only and has a regular pattern, the waveform has also. For a more usual scene the waveform might be as in (iii) containing not only black and white but also many shades in between. The whole line is drawn in 64 μs.

7.4.1 Add Colour

Things get more complicated when colour is added. Fortunately any colour can be made up by mixing together three *primary* colours only: red, green and blue (R, G, B). Briefly, the camera separates light from the scene into these primaries and directs each to a special pick-up tube. The electrical outputs from the tubes are mixed in such a way that they can be separated later and the mixture forms the *chrominance* (colour) waveform. This is transmitted together with the synchronizing pulses and the luminance.

In the receiving CRT there are now three electron guns, one for each primary colour. The screen fluorescent coating is more complicated than for monochrome. It is coated with microscopic phosphor dots in groups of three. When any dot of a group is hit by an electron beam it glows in its own colour. A special perforated mask ensures that electrons from each gun strike the appropriate phosphor dot, e.g. those from the electron gun for red are directed only to the phosphor dots which glow red. Each phosphor dot in a group therefore contributes an amount of its colour as determined by the strength of its electron beam. As an example, illuminating the red and green phosphor dots but not the blue in a group produces yellow. Add various small amounts of blue to obtain a range of pastel greens (faintly exciting the green dot on its own gives a pastel green but of the basic shade only).

It is a tribute to modern engineering that colour tv is capable of such exquisite reproduction – yet in Section 7.6 we find that improvements are on their way!

7.5 Bandwidth

The minimum bandwidth required to carry a colour tv programme satisfactorily has been arrived at via a mixture of theory and practice. The total tv signal contains the luminance and chrominance information, line and frame synchronizing pulses, a *colour burst* and the sound, quite a mixture. The colour burst follows each line synchronizing pulse and its purpose is to keep an oscillator in the colour section

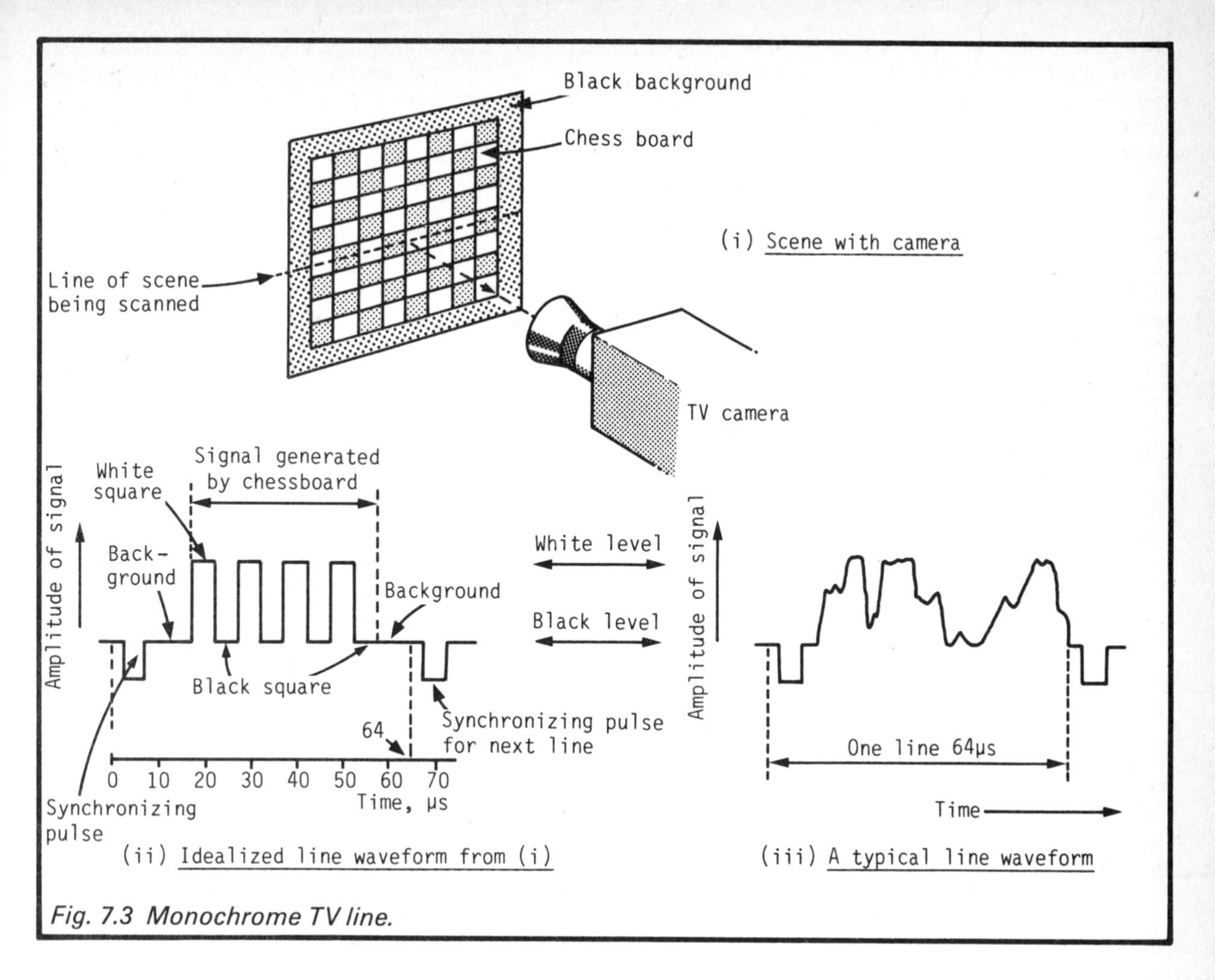

Fig. 7.3 Monochrome TV line.

of the receiver exactly at the right frequency. Taking a 625-line system, the inference is that for a picture of this number of lines each picture element must be about 1/600 of the picture height (allowing for the fact that not all lines are effective). Assuming that the width of a picture element is the same as its height and with a screen width 4/3 times the height, there are therefore 600 x 4/3 elements per line. The total number of picture elements per frame is therefore 600 x 600 x 4/3 and for 25 complete frames per second:

number of picture elements per second =

$$600 \times 600 \times \frac{4}{3} \times 25 = 12 \text{ million.}$$

Because every picture element tells a story of both brightness and colour, an enormous amount of information is needed to define a tv picture and its sound. In fact the overall bandwidth required, which is a compromise between perfection and economics is some 6–7 MHz. This can be compared with that for high quality sound at about 15 kHz, i.e. 400 times as much or for telephony at around 3 kHz, 2000 times as much. A simple comparison but it explains why it is possible for a satellite channel to carry hundreds of telephone circuits but perhaps only one or two for tv.

7.6 The New Look

Whereas most viewers would regard the existing tv reception as good enough, the more fastidious may not be so satisfied. The present colour tv standards have been with us for some 30 years and most homes have receivers designed to them. The introduction of change is therefore not a simple matter. However technology has advanced considerably during this period and the advent of DBS gives Europe the chance to introduce improvements.

The main defects in the older style tv transmissions arise from the way the video signal is put together. Perhaps the most noticeable are due to *cross colour* and *cross luminance*. The first shows itself as spurious swirling coloured patterns over parts of the picture containing fine detail such as narrow stripes on clothing. Although most viewers have become accustomed to the effect and ignore it, the picture is not a faithful reproduction and would be improved without it. Cross luminance shows itself as variations in brightness where extreme colour changes occur, resulting in tiny shimmering dots along the line of colour change, especially at vertical boundaries.

These difficulties arise from the fact that the luminance and chrominance signals share the same frequency band, in a way therefore they are mixed together. In the tv receiver these components have to be separated but with the PAL system this cannot be done completely, hence the degradation of picture quality. (PAL stands for Phase-Alternating-Line, a concept we need not grapple with.)

Considerable research has been undertaken by more than one organization to improve the picture quality and the system considered to be technically superior and generally accepted for Europe was brought to fruition in the early 1980's by the UK Independent Broadcasting Authority (IBA). The main change proposed is that the video signal should no longer contain a frequency mixture of luminance and chrominance information but that these should be separated in time. Thus for each tv line the two arrive in "packets" sent one after the other. The system is known as "Multiplexed Analogue Components" (MAC), in fact describing just what it does, the word "multiplex" indicating the transmission of several *separate* elements. Graphically the system is best illustrated on a base of time as in Figure 7.4.

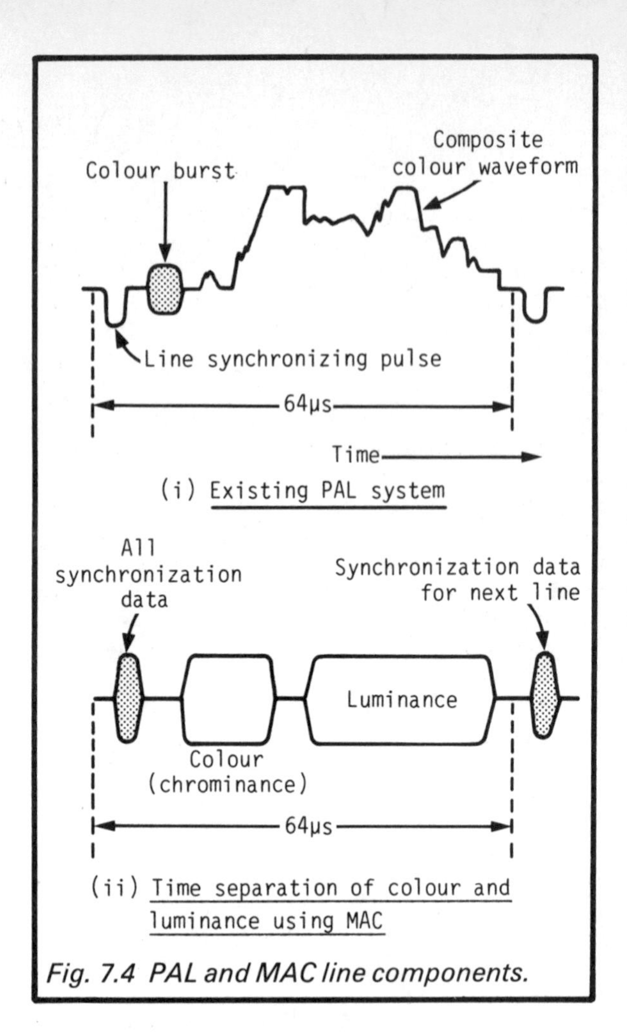

Fig. 7.4 PAL and MAC line components.

Figure 7.4(i) is a similar example to Figure 7.3(iii) but now with the colour burst and a more realistic shape of the synchronizing pulses – nothing changes in no time. With this PAL line signal the waveform at any instant after the synchronizing pulse and colour burst represents the total amplitude of the luminance and chrominance frequencies which control the spot. Moving on to the MAC line signal in Figure 7.4(ii) it is evident that luminance and chrominance cannot mix because they arrive at different times. Picture quality is therefore enhanced compared with PAL because a composite signal does not have to be unravelled first. System noise is also reduced. Additional circuit complexities are necessary to achieve the result but the technology is now with us. As an example, the chrominance for a line must be held while awaiting the arrival of the luminance, only then can the line be drawn by the spot.

Sound has not been included in this explanation but the MAC system has this in mind in that improvements in sound quality and stereo are also possible on future DBS systems. The sound system is indicated in a code preceding the letters MAC. For Europe C-MAC and D-MAC are in greatest favour and with either of these there will be "digital" sound with the enhanced fidelity which is now obtainable from compact discs.

Arrival of MAC does not immediately make all existing tv sets obsolete. Whatever system the channel broadcasts on and whatever the tv set in use, adapters fitted between the satellite receiver and the tv set will be available to make the required changes. Alternatively the extra circuitry may be included in the receiver itself.

There is also much talk in technical circles of "wide", "large-screen" and "high definition" tv (say 40" x 24" with over 1000 lines). It is unlikely that wider frequency bands than the present 27 MHz will be allowed so any enhanced features will probably be achieved through additional signal processing as in the case of MAC. The MAC system is much better suited to this than PAL hence we can look forward to even more dazzling pictures, perhaps ultimately without the "box" for the much delayed flat screen (i.e. with no lengthy tube behind it) must surely arrive one day.

Chapter 8

THE DOWNLINK

Study of the downlink in all its parts gains the greatest intimacy with satellite tv principles. The most dominant feature of the downlink on Earth is the dish for it is undisguised and in full view. It is also the one item about which there is much argument and guesswork, hence, although only a small part of the downlink, it is worthy of special consideration. Both satellites and Earth installations employ dish antennas, constructed on the same basic principles. Up above a dish sends out the radio wave, down below a dish collects it so it is appropriate that we understand the dish and its characteristics first. By so doing it is possible to eliminate much of the uncertainty in deciding the dish size for a particular installation.

Subsequently the downlink is examined as three separate parts, the satellite, the receiving antenna and the space in between. Putting these together gives the whole downlink story. Incidentally when experts refer to a *downlink budget*, it is in terms of decibels, not money.

All the formulae required are in Appendix 8 together with helpful Tables. From these quick estimates can be made. Right from the start however we must remember that the figures obtained are not precise, nor can they be considering the number of variables around, some of which are Nature's and beyond human control. Nevertheless our calculations form a good guide and practical examples using them are developed as we go.

8.1 The TVRO Dish

The aim of any antenna is to send out or gather in as much radio signal as possible and in the case of the dish the larger it is, the better it does the job. Yet for the home viewer large dishes are cumbersome, unwieldy, difficult to mount and definitely not things of beauty. Moreover installation is not as straightforward as that for a normal tv terrestrial antenna which is usually no more than a cluster of short aluminium rods. Hence for the receiving dish a compromise must be chosen between size and signal pick-up.

In carrying out its function of collecting energy from an oncoming radio wave at such high frequencies, the receiving dish intercepts the wave and directs it to a single collecting point. Many of us have used the same principle in our younger days when we were perhaps proud possessors of a *burning glass.* This is a lens used to collect the Sun's rays and focus them into a point of bright light and heat onto a piece of paper which with luck then ignites. The same principle is also extended to the concave mirror which has the ability to focus light rays to a point in front of it. Light is reflected by a mirrored surface, radio waves are similarly reflected by a metal surface, the rules governing the reflection being the same in both cases as shown in Figure 8.1(i). Here, for the convenience of illustration we use the standard way of representing light and radio waves as "rays" with arrows indicating direction of travel.

Next consider a *spherical* dish, i.e. one which is a small part of the shell of a hollow sphere. Most of the energy of a radio wave it intercepts will be gathered in and reflected towards a point known as the focus [Fig.8.1(ii)]. The word "most" is used because a simple dish such as this suffers from *spherical aberration* (straying from the path) in that energy reflected from around the edge of the dish does not converge accurately on the focus, as shown in (iii) of the Figure. Hence although dish antennas appear to be spherical, they are not, they are in fact *paraboloid*, a slightly different shape.

8.1.1 Profile

A parabola is a particular form of *conic section* in that the shape is revealed when a cone is sliced in a certain way. Figure 8.2(i) gives examples demonstrating how, as the angle of cut is changed, the circle, ellipse and parabola are exposed. The shapes can be defined mathematically[A8.1] and the feature of overriding importance for the parabola is that there is no spherical aberration. Note however that a parabolic dish only accurately directs incoming rays to its focus provided that they arrive parallel to the principal axis. Accordingly a dish must be set up so that this axis points directly to the satellite.

The parabolic dish therefore has exactly the characteristics required, so it is the perfect shape for both transmitting and receiving satellite antennas.

A parabola as examined in more detail in Appendix 8 is shown in Figure 8.2(ii). The actual profile in any particular case is determined by the focal length chosen.

Figure 8.3(i) indicates how a practical satellite dish is constructed. This one is referred to as a *primary focus* type and at the focus is the LNB (Sect.8.1.3) to which the wave energy is directed. The LNB is cabled to the receiving equipment within the premises. The primary focus dish has an obvious disadvantage in that the LNB and its supports partially block the incoming wave and so effectively reduce the collecting area of the dish, especially with the smaller ones. To obviate this the *offset* dish is gaining popularity. It avoids shadowing by the LNB by mounting it lower as shown in Figure 8.3(ii). The profile of the dish is modified accordingly.

There is a third type more likely to be used for the smaller dishes. It is constructed similarly to Figure 8.3(i) except that at the apex of the LNB tripod there

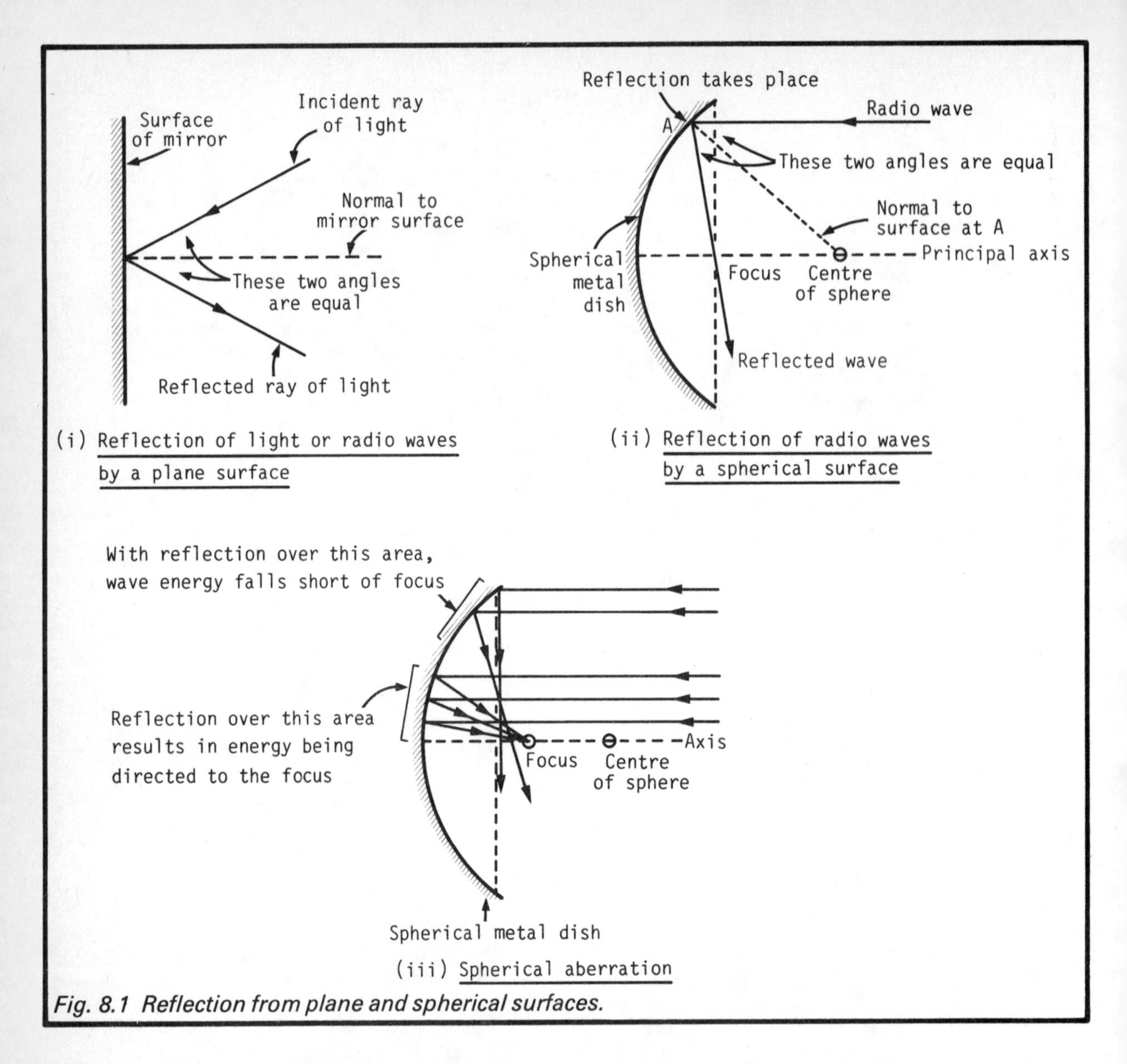

Fig. 8.1 Reflection from plane and spherical surfaces.

is instead a *hyperbolic subreflector* (a hyperbolic surface is like parabolic but turned inside out). This directs the wave through a hole in the centre of the main dish onto the LNB directly behind. The principle is known as *Cassegrain* with rear feed and the wave paths are as shown in Figure A10.1.

The question of size of dish, usually in terms of how small a one is feasible, needs a knowledge of both strength of signal and also of interfering noise (Sect.2.7) for any particular satellite and ground location. This is discussed in depth later in the Chapter.

8.1.2 Siting and Alignment

Section 8.1.1 indicates the importance of aligning a dish antenna so that its axis points exactly to the satellite. There must also be an uninterrupted path between the two otherwise the oncoming radio wave will be blocked. Even passing aircraft can cause "picture flutter". A dish must therefore be sited so that, except for the inevitable clouds, rain or mist, there is an unobstructed "view" of the satellite, not only from the centre of the dish but also from around the edge. For many would-be European viewers with a clear view from home or garden to around the South this presents no problem. On the other hand, for others with buildings or trees in the way things are not so simple with the larger dishes unless a suitable flat roof is available. For good reception a dish should be aimed to within about half of one degree so must be bolted down on a solid base or concrete blocks. There is no question of it being able to move, let alone blow over in a strong wind.

Smaller dishes are, of course, more easily dealt with. A dish of diameter 90 cm or less can be mounted on a wall or on a suitable chimney stack (without need of planning permission – Sect.1.3.2). The smallest of all, say, 30–35 cm can even be located on a window-

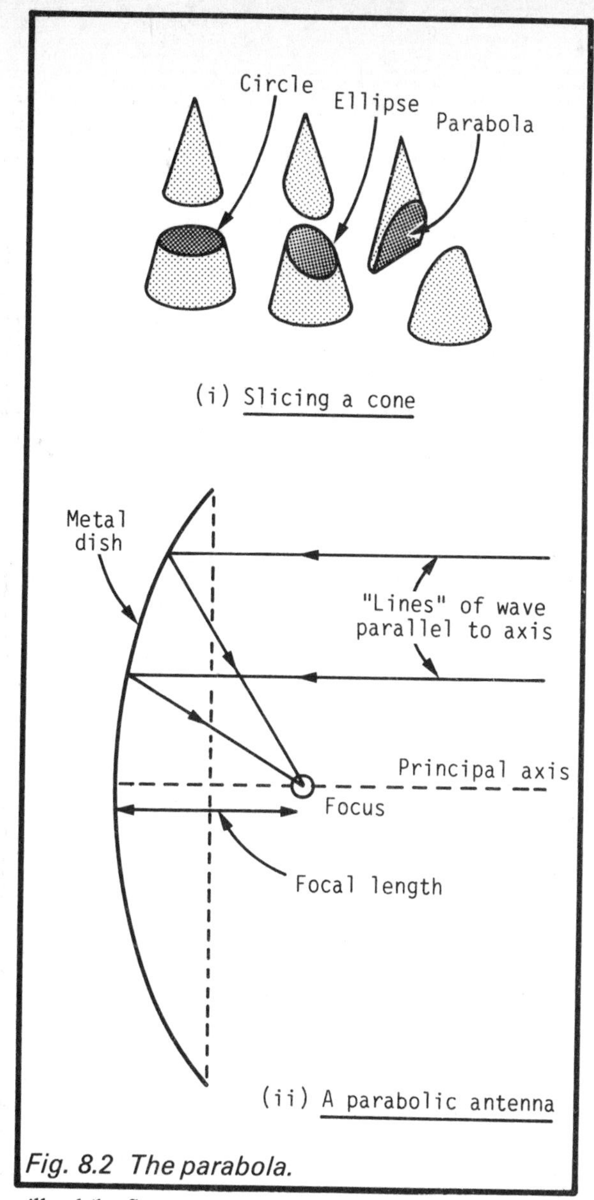

Fig. 8.2 The parabola.

sill while flat antennas are probably beset with less difficulties still.

Pointing the dish in the right direction firstly requires a knowledge of the appropriate angles of *azimuth* and *elevation* for the particular satellite. Elevation is easily understood by imagining that the satellite can be seen through a telescope. The angle the telescope makes with the horizontal (as indicated by a spirit level) is the elevation. That the elevation varies with latitude is demonstrated by Figure 8.4(i) where it is seen to be 90° at the equator (satellite directly overhead) decreasing as latitude increases. Ultimately reception becomes difficult because the radio beam arrives at such a low angle that it is obstructed by hills, buildings or even trees. At even higher latitudes (above about 81°) reception from geostationary satellites is impossible. Elevation also varies with longitude but in a more complex way which is not so easily understood from a simple sketch.

Azimuth can get quite complicated too but we can take it as the horizontal angle between true North and the satellite, measured clockwise. Imagine we are somewhere up there looking down on the Northern hemisphere as shown in Figure 8.4(ii). Then the azimuths of the two satellites from location P are as shown. The "true" reading is relative to North but azimuths are also frequently expressed as East or West (of South). Thus the azimuth of satellite 2 from P can be stated either as 150° true or as 30°E. Generally the true reading is to be preferred because most calculations work from this.

A dish is aligned by tilting it upwards from the horizontal by the angle of elevation and rotating it from a direction pointing North by the angle of azimuth. For a given satellite it is now clear that elevation and azimuth vary according to the receiving location.

Derivation of the formulae for calculation of these two important angles is from straightforward trigonometry, but messy. It is not recommended for readers who wish to sleep at night. The formulae are given in Appendix A8.2 which also includes a BASIC program for the computer enthusiasts and from these Tables A8.1 to A8.4 give a range of figures as a guide. The Tables cover most European locations and the most likely range of present and future usable satellites.

There are four variables concerned in the calculation of azimuth and elevation, i.e. the latitude and longitude of both receiving station and satellite. This makes things difficult for the production of tables or graphs which normally accommodate an input of only two. Happily the latitude of the satellite is always 0° and as Section A8.2 shows, the formulae can be developed in terms of a single variable, the *longitude difference* between satellite and receiver. One variable therefore replaces two. The price to be paid is that the difference has to be calculated in advance, a simple task however.

Note that in setting up a dish it is only necessary to point it to somewhere near the satellite so that a picture can be recognized. Then final adjustments are made by watching a tv screen (see Sect.9.2.3). There is no requirement therefore for trying to set up a dish to fractions of a degree. The values in the Tables are quoted to one decimal place for use in other calculations, here we can round them to the nearest degree without disquiet.

For any location therefore first look up its latitude and longitude. For cities and major towns the figures are normally quoted in a good atlas, otherwise they may be estimated from a map. The longitude of the satellite is also required, this will be found in satellite and allied magazines.[A1.1] We work in decimal fractions of degrees but values quoted in atlases are usually in minutes. These are changed to decimal by dividing by 60 or alternatively, if our arithmetic has seen better days, Figure 8.5 can be used. For example, 61°30′ becomes 61.5° while 61°.55′ changes to 61.9°.

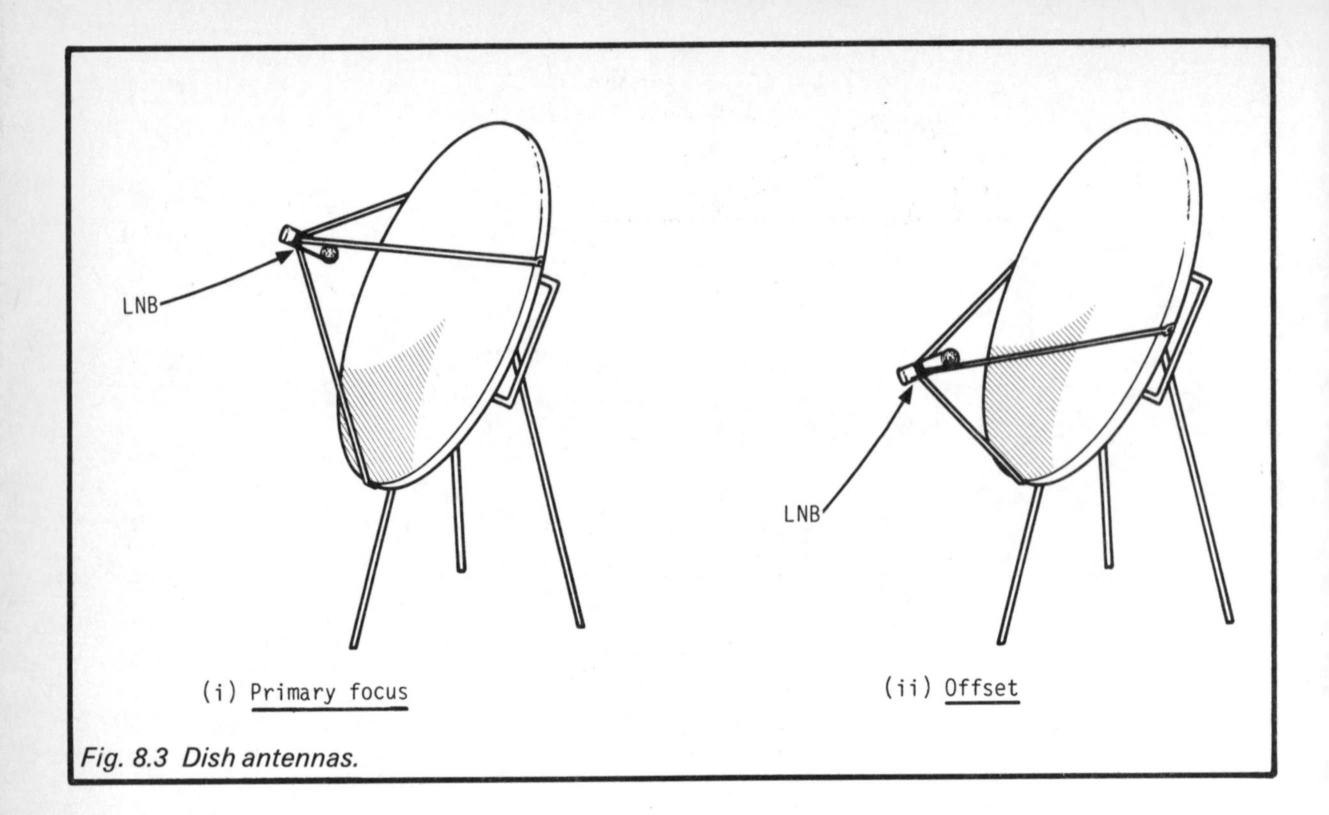

Fig. 8.3 Dish antennas.

We are dealing with angles and use a representation of them by Greek letters, θ (theta) for latitude and ϕ (phi) for longitude with subscripts r for receiving station, s for satellite and d for difference. The difference in longitude, ϕ_d between satellite and Earth station is therefore:

$$\phi_d = \phi_s - \phi_r$$

A complication is that the formulae require that longitude values to the east of 0° are considered negative. Several examples follow to show how it is done. Accordingly we can quickly get a good idea to within a degree or so of the azimuth (*Az*) and elevation (*El*) angles for any place in most of Europe. For those more daring, equations A8(6) and A8(7) enable calculations to be made for anywhere in the World.

Consider first Figure 8.6 (which is an extension of Figure 8.4) for a better idea of the azimuth calculation. Two receiving stations R1 and R2 are shown. Suppose it is required to calculate the azimuth angle at which the dish has to be set at R1 to receive satellite 1. Then:

latitude of R1 $= \theta_r = 45°$

longitude of R1 $= \phi_r = 45°$

longitude of satellite 1 $= \phi_s = 15°$

hence, longitude difference, ϕ_d, is equal to

$$\phi_s - \phi_r = 15 - 45 = -30°$$

The appropriate Table in Appendix 8 is A8.2 and at 45° latitude, with $\phi_d = -30$ is the value 140.8°. This is the required azimuth angle which is measured from the line joining R1 to the N. Pole.

The elevation to which the dish should be set is to be found in the same place but in Table A8.4, i.e. 30.3°.

Next suppose it is desired to realign the dish at R1 to receive satellite 2,

$$\phi_r = 45° \qquad \phi_s = -10°$$

(remembering that ϕ_s is east of 0° and therefore negative)

$$\therefore \quad \phi_d = \phi_s - \phi_r = -10 - 45 = -55°$$

and Table A8.2 gives for latitude 45°, ϕ_d −55°, 116.3° for *Az* and Table A8.4 gives for latitude 45°, ϕ_d −55°, 15.5° for El.

Azimuth angles for the two satellites from location R2 are also given in Figure 8.6. However for more realistic practice let us consider a few actual locations and satellites, the latter being two of the forerunners, INTELSAT V at 27.5°W and EUTELSAT ECS F1 at 13°E.

LONDON:

$$\theta_r = 51°.32' \qquad \phi_r = 0°.5'W$$

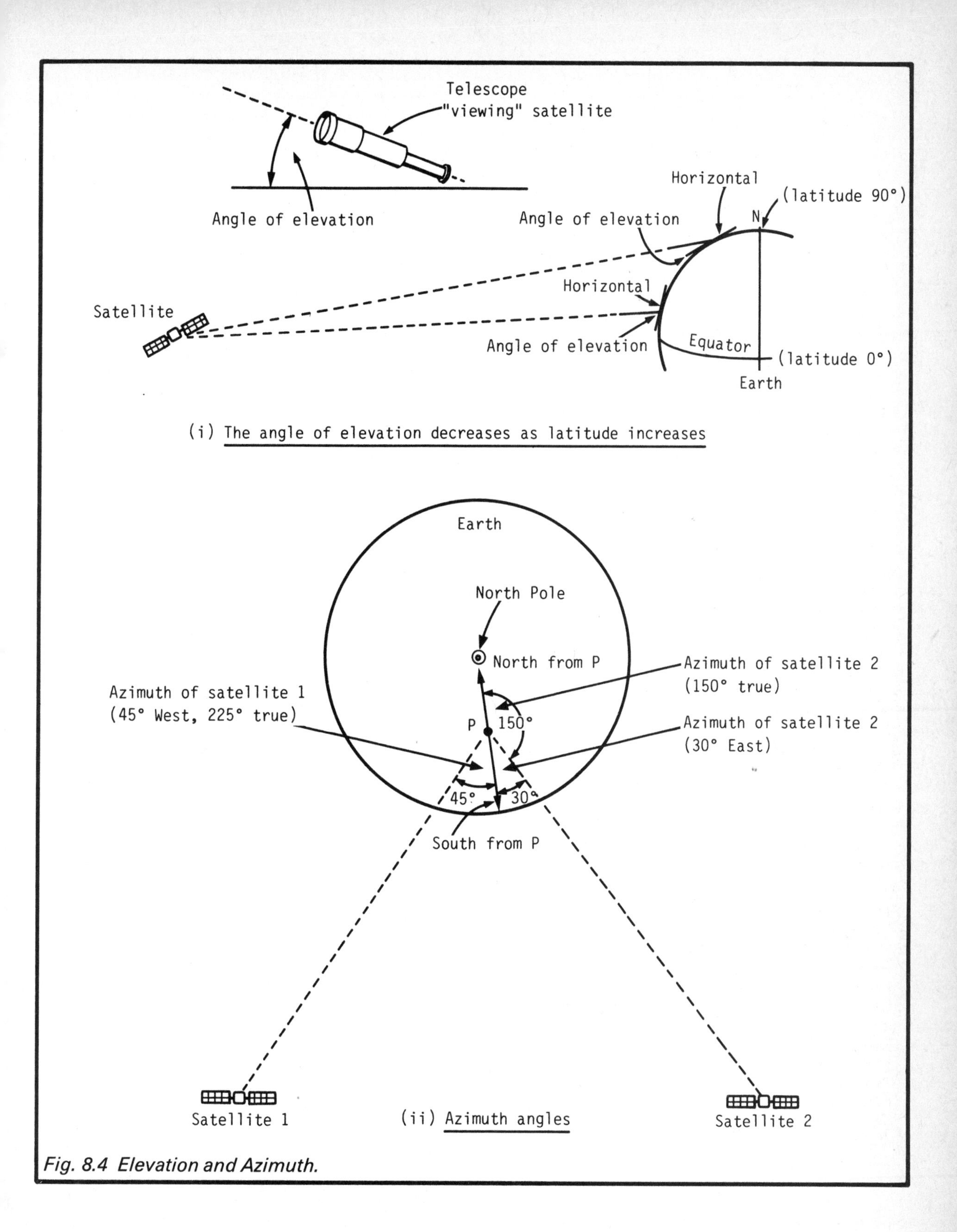

Fig. 8.4 Elevation and Azimuth.

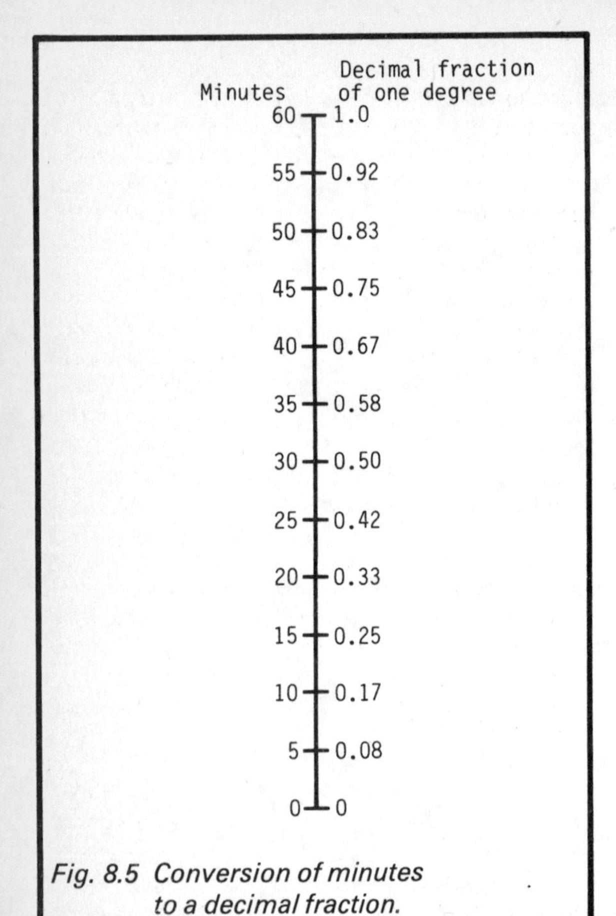

Fig. 8.5 Conversion of minutes to a decimal fraction.

Changing to decimal (Fig.8.5):

$\theta_r = 51.5°$ $\phi_r = 0.1°$

To receive INTELSAT V:

$\phi_s = 27.5°$

Then $\phi_d = \phi_s - \phi_r = 27.5 - 0.1 = 27.4°$

From Tables A8.1 and A8.3 for $\theta_r = 51$ and $\phi_d = 27.4$:

$\underline{Az = 214° \text{ (true)}}$ $\underline{El = 26°}$

(Note that for $\phi_d = 27.4$ we have to estimate from the two values for 25 and 30.)

To receive EUTELSAT ECS F1:

$\phi_s = -13°$ $\therefore \phi_d = -13 - 0.1 = -13.1$

From Tables A8.2 and A8.4,

$\underline{Az = 164°}$ $\underline{El = 30°}$

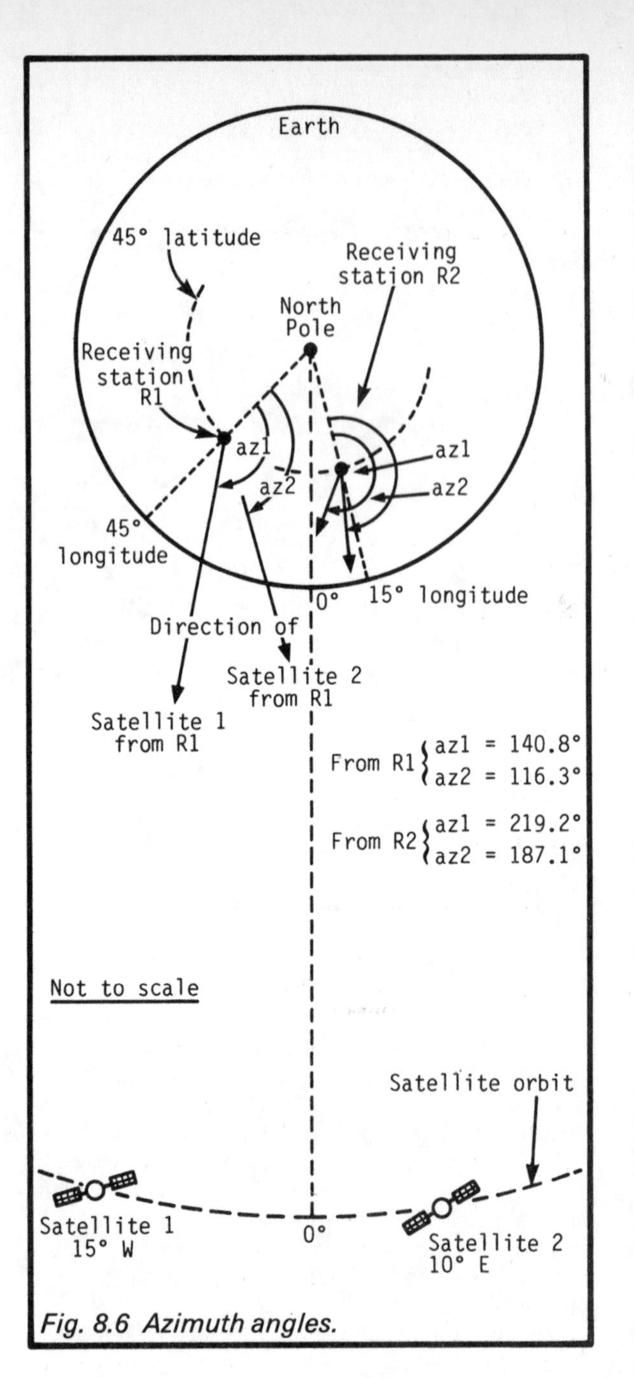

Fig. 8.6 Azimuth angles.

AMSTERDAM:

$\theta_r = 52°.22' = 52.4°$

$\phi_r = 4°.53'\text{E} = -4.9°$

To receive INTELSAT V:

$\phi_d = 27.5 - -4.9 = 32.4$

(remembering that minus minus = plus). From Tables A8.1 and A8.3,

$\underline{Az = 219°}$ $\underline{El = 23°}$

To receive EUTELSAT ECS F1:

$$\phi_d = -13 - -4.9 = -8.1$$

From Tables A8.2 and A8.4:

$\underline{Az = 170°}$ $\underline{El = 30°}$

MADRID:

$$\theta_r = 40°.24' = 40.4°$$

$$\phi_r = 3°.41' = 3.7°$$

To receive INTELSAT V:

$$\phi_d = 27.5 - 3.7 = 23.8$$

From Tables A8.1 and A8.3:

$\underline{Az = 214°}$ $\underline{El = 37°}$

To receive EUTELSAT ECS F1:

$$\phi_d = -13 - 3.7 = -16.7$$

From Tables A8.2 and A8.4:

$\underline{Az = 155°}$ $\underline{El = 40°}$.

The very last estimation for elevation may have caused some misgivings for we have θ_r at 40.4 and ϕ_d at -16.7, both inconvenient "in between" values in Table A8.4. If something better than a wild guess is required, the problem is resolved by first interpolating for $\theta_r = 40$, with the result, say 40.6, then for $\theta_r = 41$, say 39.6. Clearly about half way between these two, the result for $\theta_r = 40.4$ which, when rounded to the nearest whole number, is 40. More sophisticated methods can of course be used, even to plotting on graph paper or use of formula but here we are looking for a guide only, not a precise answer.

We conclude this Section with the thought that already those readers who are unsure that the requirement of an uninterrupted view of the satellite can be met, are now able to reach a decision. All that is required is that they know where North is and are able to estimate angles, with the help if necessary, of a compass and a protractor. More on this subject later.

8.1.3 The LNB

The Low Noise Blockconverter is the outdoor unit associated with the dish. It is probably the most important item of equipment in the receiving installation for without an efficient LNB the whole system is at risk. It is placed right at the focus of a home dish to avoid the extra expense and inconvenience of a long length of waveguide (Sect.3.5). It has two functions, to accept the weak incoming signals reflected from the dish surface, then to amplify and convert them to a lower frequency (Sect.4.5) for transmission over a special cable into the home.

In Section 3.5 it is shown that radio waves can be enclosed and travel within a tube or guide. The *feed horn* of the LNB (see Figs.8.3 and 8.7) is a specially shaped device fitted to a short section of waveguide from which waves can be projected onto the dish for transmitting or equally collected from a dish when receiving. This is rather like the megaphone as a transmitter and the old-fashioned hearing trumpet as a receiver. This analogy must not be taken too far but already some similarity in the shapes may be recognized. The dimensions of the horn are controlled by the range of wavelengths with which it is to be used. In Figure 8.7(ii) the body of the LNB contains the electronics – a low-noise amplifier (a discussion of the effects of noise follows in Section 8.5) and the block converter. The latter is a frequency-changer which accepts the incoming band (or block) of signals and changes them to a similar band but centralized on a lower frequency. If this were not done, signals at around 12 GHz would need to be fed by a waveguide directly into the home. Use of cable instead is not practicable because cable losses increase with frequency and for 12 GHz, a cable would be very expensive indeed. It is therefore better to lower the frequency first so that cable can be employed. The operating characteristics of LNB's are quoted by the manufacturers and one important feature which should not be missed is the frequency range over which the LNB operates. Some used for transmissions from ASTRA, EUTELSAT and INTELSAT may cover a range of only, say, 10.95 – 11.7 GHz. For full European DBS working the range is 11.7 to 12.5 GHz so installations which have worked happily with the earlier satellites may need a change of LNB for DBS. Many manufacturers now supply equipment which is "future proofed", i.e. it already caters for or is easily adaptable for proposed future changes.

From Section 3.1.1 we note that there are four different wave polarizations, vertical, V, horizontal, H, left-hand and right-hand circular, LHC and RHC. The feedhorn of the LNB is sensitive to these and therefore the LNB must be rotated to match the polarization of the incoming wave. Rotation from this position by 90° (a quarter of a turn) will then produce minimum signal (but maximum to a signal with opposite polarization).

Alternatively feedhorns are available with a *polarizer* or *polarotor* built in. The change from one polarity to the other is effected by rotating a flexible membrane within the waveguide. It is switched into either of the two positions by an electromagnet remotely controlled from the indoor receiver. Nowadays most antennas are fitted with polarotors.

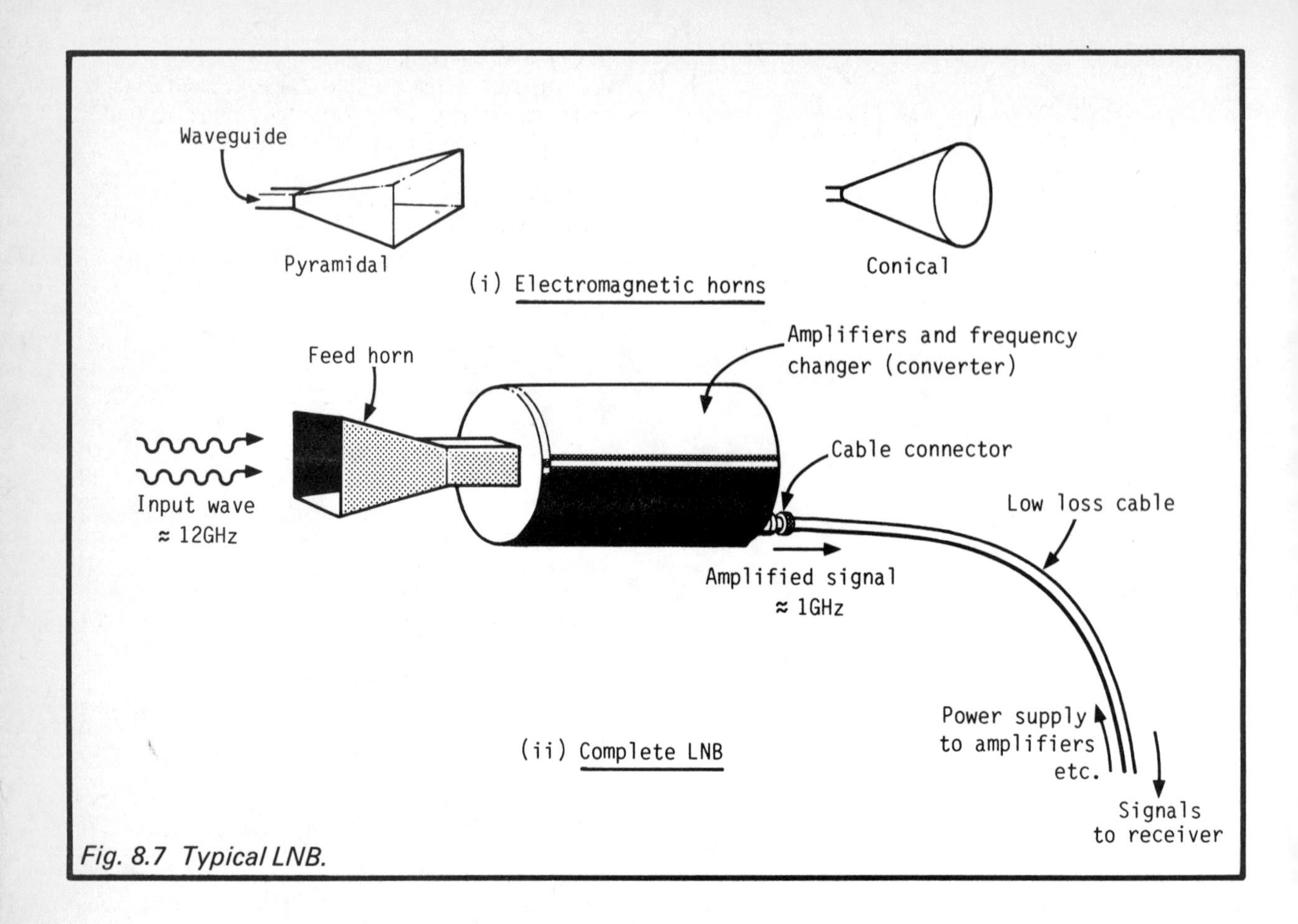

Fig. 8.7 Typical LNB.

8.1.4 Polarization Offset

Like the phenomenon from which it springs, *polarization offset* is not something which rapidly becomes crystal clear. In fact getting to grips with the concept is probably our most difficult undertaking. However polarization offset figures are frequently to be found with those for azimuth and elevation because they do affect the LNB or the dish as a whole. It is therefore desirable that we should have some knowledge of how they arise and are used.

The *propagation plane* of a wave is that which contains both the direction of propagation and the direction of electric vibrations. When facing an incoming vertically polarized wave therefore, the polarization plane might be imagined as a sheet of paper held vertically and end-on. Figure 8.8 attempts to show how a polarization angle arises when the polarization planes of satellite and receiving dish are not in line. The drawing must not be taken too literally otherwise it can lead to wrong conclusions. A single diagram such as this cannot show how one plane is rotated relative to the other, hence it must be accepted as an idea only of the planes and how they can become mismatched. Note from the drawing that the polarization plane of the wave emitted from a satellite antenna cuts through the centre of the footprint and that if the receiving antenna is in this position the two antenna polarization planes are in line – no problem. On the other hand if the receiving antenna is in any other position the Figure indicates that the two polarization planes are at an angle. This angle is known as the *polarization angle* and the greater it is, the lower the signal pick-up. However, by rotating the receiving dish or LNB the effect can be cancelled out. The amount of rotation is called the *polarization offset.* The figures are calculated as in Section A8.3 and the computer program developed for calculation of azimuth and elevation (A8.2) includes these also. The figures for Europe are given in Table A8.5. These are to the nearest degree and apply whether ϕ_d is positive or negative (Sect.8.1.2).

To use the Table, first calculate ϕ_d, look up the polarization offset value, then, standing behind the dish and looking towards the satellite:

> rotation should be clockwise if the satellite lies to the West (ϕ_d positive)
>
> rotation should be anticlockwise if the satellite lies to the East (ϕ_d negative).

In practice, where there is remote control of the LNB for changing the polarization there will also be a "skew" control for fine tuning. This control usually

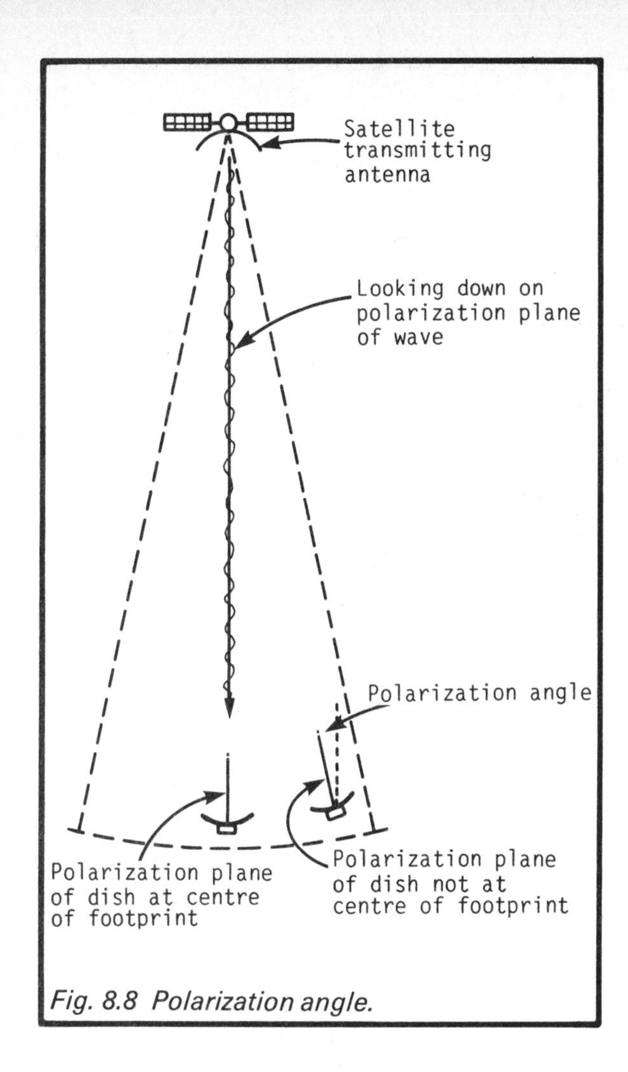

Fig. 8.8 Polarization angle.

sorts out all the polarization discrepancies together including the effects of depolarization on linear waves mentioned in Section 3.1.1. The polarization offset is therefore automatically taken into account.

8.1.5 Squaring the Circle

The aim of the parabolic dish is to collect the radio wave and focus its power into a single pick-up point. It is also possible to make up a flat array of relatively tiny antenna elements of a different type altogether. At these high frequencies a slot or round hole of a certain size as determined by the wavelength (see Section 3.5 on waveguides) and cut in a conducting surface acts as an antenna. The power intercepted by such an element is naturally very small but many can be connected together in such a way that the powers add up and a feedhorn is not therefore required. Technically the principle is known as the *phased array*. There might be as many as 500 – 1000 of these miniature antennas cut into a flat plate to form a single satellite receiving antenna.

The technique is relatively new but already flat antennas are able to compete technically with parabolic dishes and although their efficiencies are at present at the lower end of the scale, there is little doubt that there will be improvements in the future. For the same antenna gain the area required is slightly larger than for the dish. The LNB is attached flush to the back of the plate.

8.2 Antenna Evaluation

One of the complexities of communication engineering is that of assessing how well an antenna does its job. The only practical way is by comparison with a known, easily reproduced standard. This is done by rating the antenna against a hypothetical elementary one supposedly radiating from or receiving at a single point and equally in all directions. Such an antenna cannot be constructed. This does not matter because its performance can be calculated instead. It will be seen how useful such a concept is for not only can dish antennas be evaluated by this method but also any other type.

We ourselves are mainly concerned with the receiving antenna but theoretical explanations are often simplified by considering a transmitting one instead. It is a fact that the properties of an antenna used as a radiator are similar in many respects to those when it is used for reception. Such a reciprocal relationship enables us to produce explanations more easily by switching between transmitting and receiving as the occasion demands. This in no way detracts from the final conclusions.

8.2.1 The Isotropic Antenna

The elementary antenna is known as *isotropic* (from Greek, *iso*, equal and *tropos*, turn, i.e. having the same properties in all directions).

Consider a transmitting antenna first and imagine it to be placed in the centre of a sphere as shown in Figure 8.9. The Figure shows how the formula for power flux density (pfd) is developed (i.e. how much signal power exists over a given area). It is an indication of the *strength* of the signal. The conclusion is that if an isotropic antenna is fed with a signal power of P_t watts, then the pfd at a distance away of r metres is

$$\frac{P_t}{4\pi r^2} \text{ watts per square metre (W/m}^2\text{)}.$$

Hence, for example, a 100W transmitter (the power consumed by a household lamp) feeding an isotropic antenna would result in a signal power flux density at only 1 km away of:

$$\frac{100}{4\pi \times (1000)^2} \text{ W/m}^2$$

i.e. about 8 μW/m^2 (microwatts per square metre – one microwatt is one millionth of one watt). We keep this in mind to see later how much better a parabolic dish can be.

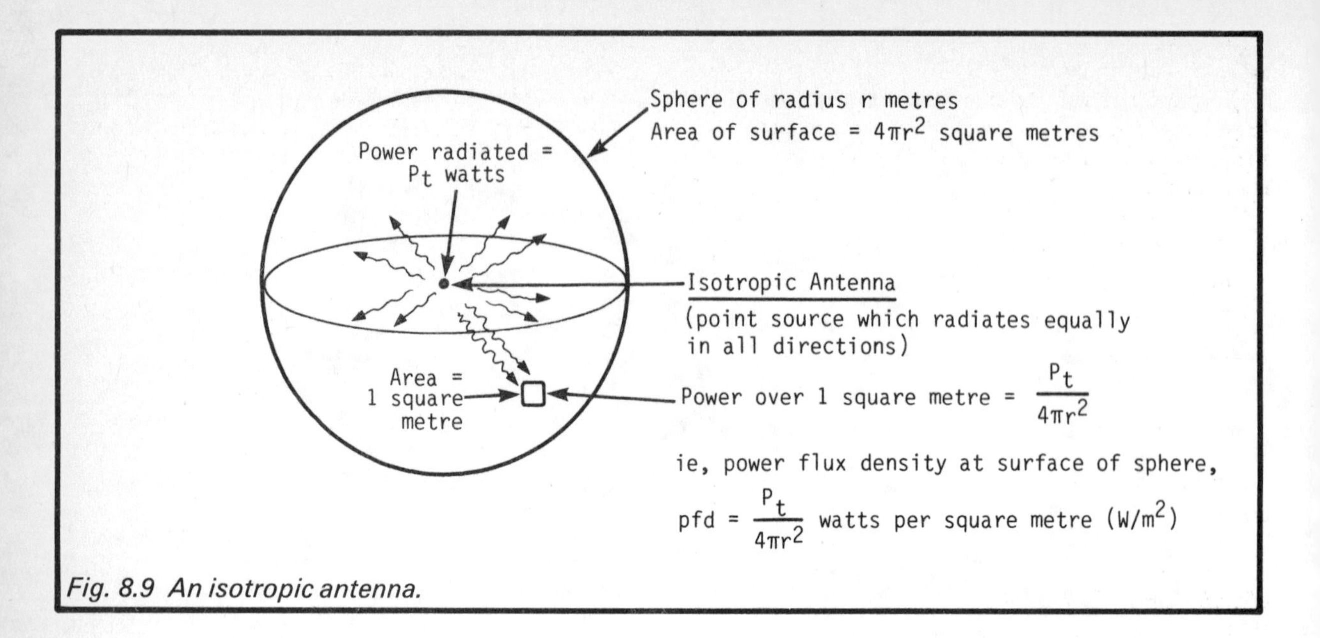

Fig. 8.9 An isotropic antenna.

8.2.2 Antenna Gain

The goodness of a practical antenna is summed up fairly well by its *gain*. This is a figure expressing how well it radiates or receives energy in or from a given direction compared with the isotropic. Because generally practical antennas are expected to be better, the gain G is positive. The whole story is given in Appendix 8 (Sect.A8.4) together with a computer program for quick calculation. The formula for G shows that for a dish, the gain varies directly both with dish diameter and frequency. [Equation A8(16)].

A transmitting dish antenna is designed to concentrate as much power as possible into a pencil-like beam. Figure 8.10 shows that for this purpose, as far as the area on Earth served by the beam is concerned, most of the power radiated from an isotropic antenna is wasted. Only a tiny proportion is radiated in the desired direction, the remainder shoots into space and is lost. Conversely with the parabolic dish most of the power is sent down the beam so the parabolic dish is said to have a high gain in the beam direction which is along its principal axis (see Fig. A8.1). In fact gains G of 10,000 are usual. Compared with the isotropic antenna in the preceding Section which theoretically would produce 8 $\mu W/m^2$ at 1 km, such an antenna would have a beam power density of 80,000 $\mu W/m^2$ at the same distance.

We must be careful not to be misled by the isotropic antenna concept. There is a quantity labelled EIRP in Section 8.3.1, it is the Effective Isotropically Radiated Power which is the antenna power input multiplied by the antenna gain. Accordingly it might seem that 100 watts power input to an antenna of $G = 10{,}000$ would result in an output power into the beam of 100 x 10,000 watts = 1000 kW (kilowatts), enough to keep 1000 one-bar electric fires going. Of course this is nonsense, no greater power can be radiated from an antenna than is put in. Always remember that such figures do not represent actual power, they are hypothetical. The

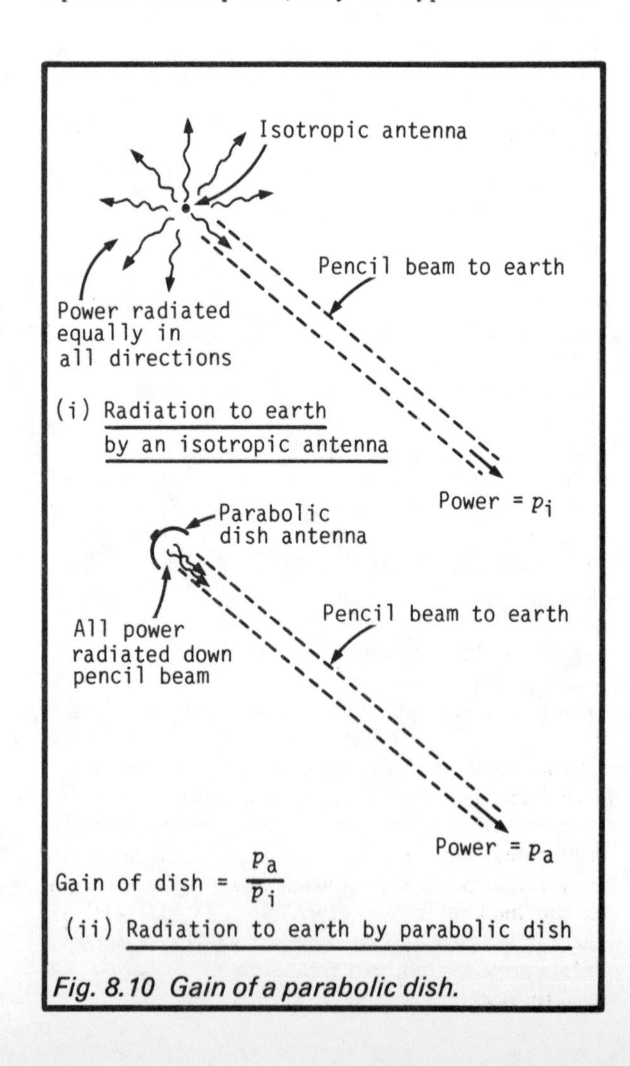

Fig. 8.10 Gain of a parabolic dish.

system will make more sense when the receiving antenna is also included.

Practical parabolic antennas are not 100% efficient for an appreciable amount of the energy is lost. In the case of a receiving dish for example there are losses due to:

(i) surface irregularities

(ii) shadowing of the reflector by the LNB, cables, struts etc.

(iii) not all the energy being intercepted at the edges

(iv) cable losses.

The Greek letter η (eta) is used as a symbol for efficiency and in practice $\eta = 0.5 - 0.6$ (50 – 70%) is common, the very best reach $\eta = 0.8$.

The area used in gain calculations is known as the *effective area* (A_{eff}), which is simply equal to ηA [Ref. A8(14)] where A is the actual area of the dish.

8.3 Link Signal Levels

Let us first recapitulate on what we are looking for. For a satisfactory picture, the input signal to a tv set must not only be of sufficient amplitude but the signal-to-noise ratio (Sect.2.7) must also be adequate. The signal arriving on Earth is weak because of the losses sustained on its long journey. Before anything else therefore we need to estimate the signal power output from the receiving dish. To do this we start with the signal power radiated by the satellite and add and subtract the gains and losses experienced on the journey down the link as far as the output of the receiving dish. This assumes a certain dish size chosen for the calculation. Fine, but what we probably prefer to know is what is the smallest dish we can employ commensurate with an adequate tv picture. To find this it is first necessary to determine what dish signal output is required for a satisfactory picture. Now this is one of the more complex considerations because as pointed out in Section 2.7, the signal-to-noise ratio is involved and this ratio changes as the signal progresses from LNB through the satellite receiver to the tv set. Moreover, not only are there several ways of defining how equipment handles a mixture of signals and noise, but also expectations of viewers vary over a wide range. So faced with such imponderables, we cheat a little by working backwards from figures published by the experts but not until a complete system for assessment of the downlink has been developed.

It is better to use decibels now to avoid being confronted with multiplication of ratios and unmanageable numbers. Hence readers unaccustomed to the method may wish to have another look at Section 2.5 first. Those who have not met decibels before may rest assured that generally we soon get used to them.

8.3.1 Power in the Sky

We touched upon EIRP in Section 8.2.2 (Effective Isotropically Radiated Power). It is a hypothetical consideration as in the case of the isotropic antenna itself. For a satellite transmitting antenna:

$$\text{eirp} = P_t \times G_t$$

where P_t is the power fed to the antenna and G_t is the antenna gain. With the reminder that

x dB indicates a power ratio

x dBW indicates a power ratio with respect to *one watt*

here is a practical example, typical of a DBS satellite:

Suppose a satellite antenna is fed with a power $P_t = 200$ W. Expressed in decibels(A2.3) this is equivalent to +23 dBW. Suppose also that the antenna has a gain over the isotropic of $G_t = 43$ dB. The satellite eirp $= P_t \times G_t = +23$ dBW $+ 43$ dB $= +66$ dBW meaning that a +23 dBW *power level* raised by a 43 dB *ratio* becomes a +66 dBW power level.

This assumes that no other losses occur within the satellite between the transponder output terminals and the transmitting antenna. When electricity flows in wires or waveguides, it creates heat just as an electric fire does. In the case of the signal the heat is infinitesimal but the power loss it creates cannot be ignored. Exactly the same effect arises in the electrical wiring in a house, the power flowing along the cables gives rise to a tiny amount of heat in them and so power is consumed. In this case the effect is unnoticeable and because there is an abundant supply of power from the generating stations it is of no consequence. With the satellite signal, things are different for the power generated is all there is. It is precious so wiring losses must be kept as low as possible.

It is generally considered that a loss of about 1 dB or so is appropriate for satellites. The eirp is therefore less than that shown above. Designating the wiring loss by L_{tw} (transmitter wiring):

$$\text{satellite eirp} = P_t + G_t - L_{tw}$$

where P_t is expressed in dBW, with G_t and L_{tw} in dB, so we have for this example:

$$+23\text{ dBW} + 43\text{ dB} - 1\text{ dB} = +65\text{ dBW}$$

A reminder once more – this is not actual power but effective isotropically radiated power.

8.3.2 Journey Through Space

Isotropic antennas throw their energy out in all directions so that which arrives at any particular location on Earth is theoretically extremely small.

In fact the calculations involved defy the ordinary calculator and only the scientific one can get the better of them. What we aim for in this Section is ultimately to determine the power in the signal when it arrives on Earth. There are two approaches to this although in fact, as Section A8.6 shows, they arise from the same fundamental formulae. The first method calculates the theoretical signal loss (or attenuation) in decibels experienced over the path between the two antennas. This is generally termed the *free space path loss*. The second method calculates the *power flux density* (pfd) in the satellite footprint. The first is developed here, the second in the next Section. It goes without saying that the two methods give the same overall answer.

Free space path loss is therefore a measure of the signal attenuation between two isotropic antennas separated by a distance d in free space. Strictly this infers a vacuum all the way so a correction for the loss in the Earth's atmosphere is added later. The isotropic antennas are themselves loss free by definition. Appendix 8 (Sect.A8.6) covers the mathematics involved arriving finally at a very useful equation:A8(19)

$$L_{fs} = 92.44 + 20(\log f + \log d) \text{ decibels (dB)}$$

where

L_{fs} is the free space loss
f is the wave frequency in GHz
d is the distance in kilometres.

(Note that d now replaces the r used earlier in considering isotropic antennas.)

A reminder for engineers is appropriate here. L_{fs} does not indicate that power is actually absorbed as the wave travels through space for there is nothing in space to do this. It simply accounts for the fact that from an isotropic antenna the electromagnetic wave spreads out in all directions, most of which is lost. Put simply, the free space loss is a measure of the power which is lost into empty space.

As yet, we only know the value of d for a receiving station immediately beneath the satellite (35786 km – Sect.5.1.2). It is also necessary to calculate satellite distance from any other location. Figure 8.4 shows that this increases with latitude. It also increases with the difference in longitude between satellite and Earth station. The formula for calculation of d therefore involves both θ_r and ϕ_d and naturally requires trigonometry.A8(17) Satellite distance is also calculated as a by-product in the computer program for azimuth and elevation (Sect.A8.2).

To try out the formula, let us calculate L_{fs} for London assuming a frequency, f, of 11.86 GHz from a satellite at 31°W. Take θ_r as 51.3°, ϕ_r as 0.1°W. Then

$$\phi_d = \phi_s - \phi_r = 31 - 0.1 = 30.9°$$

from equation A8(17)

$$d = h\sqrt{1 + 0.42(1 - \cos\theta_r \cdot \cos\phi_d)}$$

$$= 35786\sqrt{1 + 0.42(1 - 0.6252 \times 0.8581)}$$

$$= 39114 \text{ km},$$

then from equation A8(19)

$$L_{fs} = 92.44 + 20(\log 11.86 + \log 39114)$$

$$= 92.44 + 20(1.074 + 4.592)$$

$$= 205.8 \text{ dB}.$$

This is a loss so great that for normal satellite powers the signal arriving on Earth would be useless. But note that this is a theoretical base-line only, there are gains in the downlink to offset some of the loss.

Taking the range of latitudes and longitudes for Europe with ϕ_d up to 70°, then for the proposed frequency range of 11.7 to 12.5 GHz (Sect.3.4), L_{fs} varies between a minimum of 205.2 to a maximum of 206.6 dB, i.e. it is fairly constant. It is not unreasonable therefore to take a round figure of 206 dB for general work. If greater accuracy is required, there is always recourse to the formula.

The free space path loss L_{fs} does not tell the whole story for it assumes clear space and clean air. There is an additional loss to take into account due to the Earth's atmosphere and appropriately known as *atmospheric loss.*

The boundary between the atmosphere and space is not clearly defined and the effect of the various "layers" can only be expressed by a single loss value as found by measurement. There is the analogy of light from the Sun which manages to get through even when dark, heavy clouds are around, but with less brightness. Radio frequencies are also affected by diffusion and absorption in the lower layers of the atmosphere and by rain, mist and clouds.

The atmospheric loss, L_{at} obviously increases with the length of path through the atmosphere. The path length is related to the angle of elevation, i.e. the loss is greater for the lower elevations. A rough guide is given by:

elevations of 5 – 14°, $L_{at} \approx 5$ dB

elevations of 15 – 24°, $L_{at} \approx 2.5$ dB

elevations of 25 – 45°, $L_{at} \approx 1.7$ dB

As with the weather, there can be nothing precise about this. There is only a small chance that L_{at} will rise to greater values and even then by usually not more than a few decibels. But if the installation is tight on signal-to-noise ratio (usually because of a

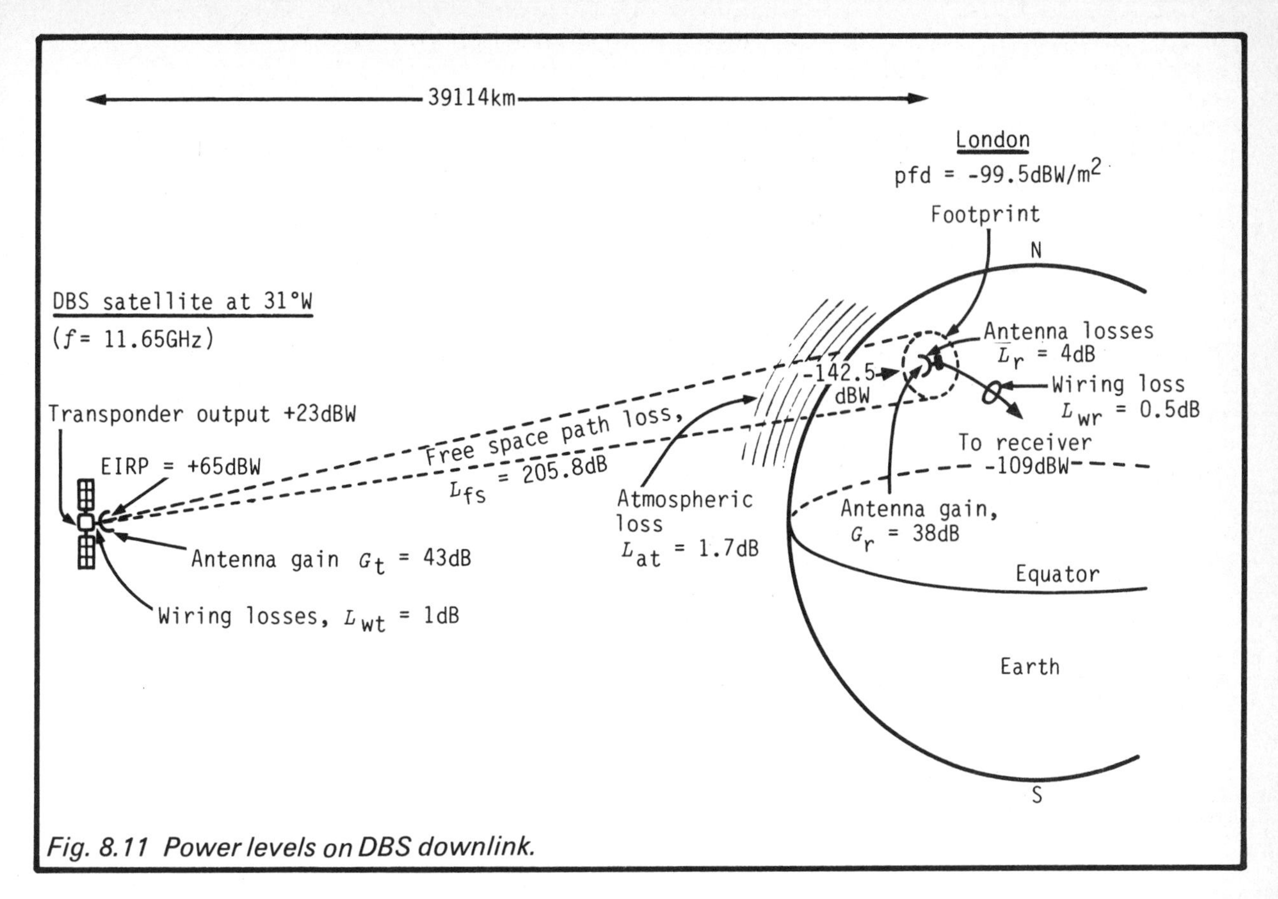

Fig. 8.11 Power levels on DBS downlink.

too small dish) it could mean that the tv picture is occasionally degraded or might even disappear with a rainstorm. Good design therefore includes an allowance for L_{at} and the greater the allowance, the less likelihood of picture loss.

We can now assess the signal power available to another isotropic antenna on Earth in or near London on the assumption that this is at the centre of the footprint. It is:

$$\text{satellite eirp} - L_{fs} - L_{at}$$

i.e. $+65\ \text{dBW} - 205.8\ \text{dB} - 1.7\ \text{dB} = -142.5\ \text{dBW}$.

The complete story is illustrated in Figure 8.11 on which the power available to the receiving dish is shown. Ignore the other data surrounding the dish at present.

8.3.3 Power on Earth

In the foregoing Section we find the signal power *level*, i.e. how much power is available to an earth-bound isotropic antenna. The alternative method of calculation is to find the *power flux density*. The pfd is measured in dBW per square metre (dBW/m^2), i.e. the amount of power available over a given area. This is in a way analogous to pressure for which, before SI units came along, we used to quote in pounds per square inch. Calculating the pfd for a given satellite is straightforward provided that the distance, d, to the area served is known. From A8(21), pfd is equal to

$$\text{satellite eirp} - 71 - 20\log d - L_{at}\ \text{dBW/m}^2$$

where d is in km, eirp in dBW and L_{at} is the atmospheric loss (dB).

Putting in the figures for our typical DBS satellite:

$$\text{pfd} = 65 - 71 - 20\log 39114 - 1.7$$

$$= 65 - 71 - 91.8 - 1.7$$

$$= \underline{-99.5\ \text{dBW/m}^2}.$$

The "power on Earth" at a receiving dish is therefore expressed either by the signal level after it has endured the free space and atmospheric losses as shown in the foregoing Section, or alternatively as the pfd, again with allowance for atmospheric loss.

It may be found that some footprint maps are labelled according to the satellite eirp. This is not so informative because they quote what the satellite transmits rather than what arrives on Earth. However, change from eirp to pfd is simply carried out as above by using Equation A8(21).

8.3.4 The Dish Delivers

The last link in the chain is the one over which we have a modicum of control. A measure of the output of a receiving antenna is obtained by:

(i) raising the input signal level by the gain, G_r ; or

(ii) multiplying the pfd by the effective area, A_{eff}

and in both cases subtracting the total losses.

The receiving antenna losses are likely to be greater than those of the satellite for the latter is kept under constant accurate control. On the ground it is unlikely that the dish will be pointing *exactly* towards the satellite, unlikely too that the LNB will be in perfect alignment with the polarization of the incoming wave. With time, dust and dirt enter the scene and experience shows that even insects sheltering in some types of LNB feedhorn create a loss. Call the total antenna losses, L_r and add an allowance for receiving antenna wiring loss L_{wr} . These might be estimated at L_r , 4 dB, L_{wr} 0.5 dB.

Consider the suggested 90 cm dish for DBS, having an efficiency at the lower end of the scale of 50%. Table A8.6 gives the gain of such a dish at, say 11.9 GHz as 38 dB [calculated from Equation A8(16)]. Accordingly the dish will deliver a signal power of:

satellite transponder power + total gains,

G − total losses, L

so moving down the link we have:

satellite transponder power +

$$G_t - L_{wt} - L_{fs} - L_{at} + G_r - L_r - L_{wr}$$

in this example:

$$+23 + 43 - 1 - 205.8 - 1.7 + 38 - 4 - 0.5$$

$$= -109 \text{ (dBW)} .$$

Alternatively we could have started from the results of Section 8.3.2 with an input signal to the dish of −142.5 dBW. Add the dish gain, G_r , subtract the losses L_r and L_{wr} to get:

$$-142.5 + 38 - 4 - 0.5 = -109 \text{ (dBW)} .$$

All the figures are shown on Figure 8.11.

The method of calculation from the pfd firstly requires a knowledge of the effective area, A_{eff} of the dish (Sect.8.2.2). Given this the signal power output follows:

signal power output =

pfd x A_{eff} minus dish losses

If A_{eff} is quoted for a particular dish, say, for the one considered above, 0.318 m^2 then,

$$\text{pfd} \times A_{eff} = -99.5 \text{ dBW/m}^2 \times 0.318 \text{ m}^2$$

$$= -104.5 \text{ dBW}$$

and subtracting the total losses of 4.5 dB gives −109 dBW as calculated above.

Putting this more plainly, we might say that a pfd of −99.5 dBW/m^2 would produce a power output of −99.5 dBW if the dish effective area were one square metre. In this case it is less, only 0.318 m^2 hence a lower output is available and this is calculated to be −104.5 dBW, i.e. 5 dB lower. Be careful of the arithmetic here, it is easy to trip up when decibels and normal ratios are mixed. Convert one way or the other e.g. because A_{eff} is a ratio we can use its decibel equivalent i.e. 10 log 0.318 = −4.975 dB. Alternatively the pfd could be changed back from decibels to a normal power.

The effective area may pose a problem. It is normally calculated by multiplying the physical dish aperture, A , by the efficiency, η (Sect.8.2.2). However few manufacturers of dish antennas seem anxious to give all their secrets away so we may have to calculate A_{eff} from the gain and efficiency.[A8(13)]

Undoubtedly for many readers this has become more than a little complicated. It does however lead to a realistic understanding of the parabolic dish. To help further, the receiving end of Figure 8.11 is expanded in Figure 8.12.

This completes the transmission story of the satellite-to-Earth downlink but for one typical full-power DBS satellite only. The figures are entered on the first line of Table 8.1. It is next profitable to add those for a working non-DBS satellite (line 2). This particular one happens to have the centre of its footprint near London so it can be used for comparison. The dish output signal powers are seen to be quite different but with them we are in a better position to discuss the most important question – what size of dish?

8.4 Dish Size

The entries in Table 8.1 give much to think about and to understand them all is progress indeed. Unfortunately there is no magic formula which decides on dish size for us.

The standard which was laid down for European DBS satellites was for a pfd of not less than −103 dBW/m^2 for 99% of the time during the worst month. This is at the edge of the primary service area and implies some 3 dB better at the centre, i.e. −100 dBW/m^2. Our own calculations for a DBS satellite as shown in the Table result in a pfd of this order (London theoretically in the footprint centre).

A sufficiently high signal-to-noise ratio must be provided for the home tv receiver to give of its best and the overall system design was such that with a

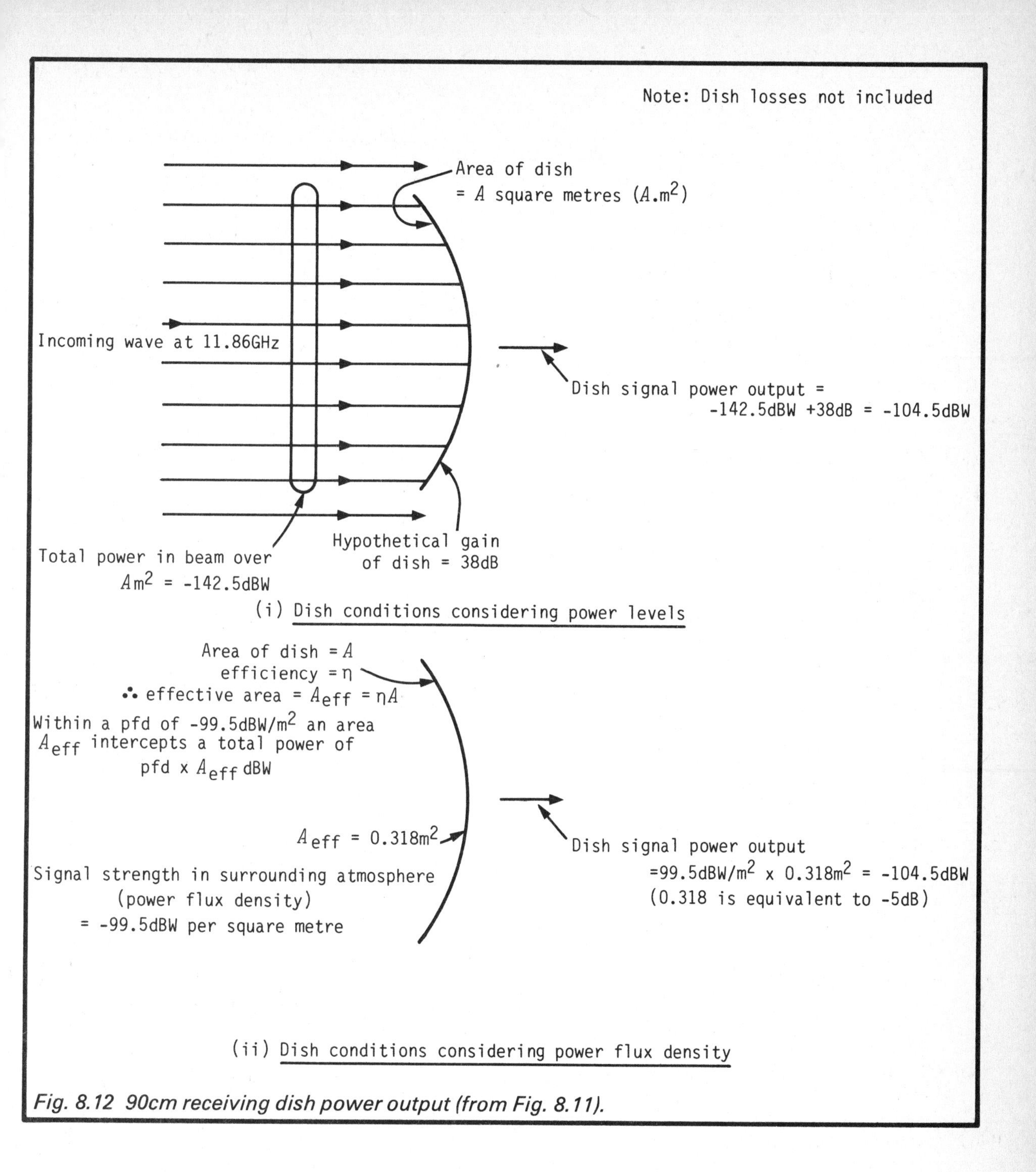

Fig. 8.12 90cm receiving dish power output (from Fig. 8.11).

pfd of -103 dBW/m^2 a 90 cm dish would be required. This was way back in 1977. Since then technical progress especially in the reduction of noise introduced by the receiving equipment, has enabled satellite power to be reduced, for example ASTRA at 45W, BSB1 at 110W (see Appendix 11). Even with lower satellite powers there is also a tendency to recommend smaller dishes than previously envisaged mainly with a view to early mass penetration, essential if high provision costs are to be recouped. With this in mind we can see what Table 8.1 has to offer.

Firstly the approximations must be kept in mind:

(i) the atmospheric loss is an average figure, it could be less for much of the time but on the other hand is liable to increase substantially for very short periods;

(ii) the dish receiving loss has been estimated at 4.5 dB. We could be lucky and have the dish and LNB aligned exactly, in which case the allowance is excessive.

TABLE 8.1 DOWNLINK SIGNAL POWERS TO LONDON (51.3°N, 0.1°W)

	SATELLITE							TRANSMISSION PATH					RECEIVING ANTENNA					
Name	Position	ϕ_d (deg)	Frequency (GHz)	Transponder output power (dBW)	Antenna gain G_t (dB)	Wiring losses L_{wt} (dB)	eirp (dBW)	Length, d (km)	Basic path loss L_{fs} (dB)	Atmospheric loss, L_{at} (dB)	Total path loss (dB)	pfd on ground (dBW/m²)	Diameter d (cm)	Efficiency η (%)	Gain, G_r (dB)	Losses, $L_r + L_{wr}$ (dB)	A_{eff} (dB)	Output signal power (dBW)
References →		A8(5)					Sect. 8.3.1	A8(17)	A8(19)	Sect. 8.3.2	Cols. 8 + 9	A8(21)			Table A8.6 or A8(15)		Table A8.7	Cols. 11–15+16 †
Column No.		1	2	3	4	5	6	7	8	9	10	11	12	13	14	15	16	17
Full DBS	31°W	30.9	11.86	+23	43	1	+65	39114	205.8	1.7	207.5	−99.5	90	50	38	4.5	−4.98	−109
EUTELSAT ECS F1 (West spot beam)	13°E	−13.1	11.65				+49	38613	205.7	1.7	207.4	−115.4	120	60	41.1	4.5	−1.7	−121.6
ASTRA*	19.2°E	−19.3	11.32				+52	38744	205.3	1.7	207.0	−112.5	65† 85	70† 60	36.2 37.8	4.5 4.5	−6.3 −4.7	−123.3 −121.7
BSB1*	31°W	30.9	11.86		42		+61	39114	205.8	1.7	207.5	−103.5	60 35	60 50	35.2 29.8	4.5 4.5	−7.7 −13.2	−115.7 −121.2

*Estimated, January 1989
†Offset antenna – figures published by manufacturer.

† Or columns 6–10+14–15

Now that satellite tv has been in operation for several years Table 8.1 can be used to compare what the DBS planners had in mind with the actual performance of working systems. The first line in the Table indicates that for full-power DBS with a 90 cm receiving dish, 50% efficiency:

max. dish signal output power = −109 dBW .

This is at the footprint centre, hence within the service area the minimum power is 3 dB lower, i.e. −112 dBW.

The second line for EUTELSAT which is not of the more powerful DBS variety shows a dish output nearly 13 dB lower. This is thought to be generally acceptable so it might appear that a fully-powered DBS satellite working to a 90 cm dish is now gilding the lily.

The lines for ASTRA and BSB1 in the Table are of special interest to UK viewers. They are intended to show how the Table is used for sorting out which satellite does what at any particular location, in our example, London. Although some of the figures are estimated, we can still get a good idea of the relative dish output signal powers. Let us compare with the EUTELSAT F1 which under all the conditions assumed, results in an output of −121.6 dBW. Only time will tell if this figure is of the right order.

The figures in Column 17 speak for themselves. With the 65 cm dish, ASTRA is about 2 dB lower. BSB1 however has more in hand at 6 dB better. It is also evident that increasing the ASTRA dish to 85 cm or reducing the BSB1 to 35 cm results in the same overall performance equal to that of the EUTELSAT. A 35 cm dish (or flat antenna equivalent) with BSB1 has the obvious advantages of low cost and small size, the latter is especially important for chimney stack and wall mounting. Note however that here we are not talking about a high power, high quality service but one which is acceptable to most viewers who require the simplest and cheapest of installations for receiving the home satellite. Little is left in hand for the inevitable rainy day and there may be some increase in sky noise because of the large beamwidth of a small dish (Sects.6.4.1 and A6.1).

Given the local pfd of any satellite, we can also assess the possibilities of satisfactory reception through a series of calculations using Table A8.7. Any practical dish may be chosen for the basic calculations then the Table quickly assesses the effect of changing to dishes of other sizes. We have seen that once the pfd on the ground is known, the dish output signal power follows from:

$$\text{pfd}\ (\text{dBW/m}^2) + A_{\text{eff}}\ (\text{dB}) - \text{total losses (dB)}$$

Note that the dish output relates directly to changes in A_{eff} since the pfd and total losses remain constant. If the dish is not circular, A_{eff} is calculated simply from $10\log(\eta A)$. As an example, with any satellite, changing from a 30 cm, 50% efficiency to a 90 cm, 60% efficiency dish improves reception by $(-4.2 - -14.5) = 10.3$ dB.

The technique is especially useful when attempting to assess the prospects of tuning in to satellites not directly intended for us. Take, for example, a far away satellite for which the pfd is quoted as −120 dBW/m^2. But firstly let us be sure of our standard. On the assumption that the dish output signal from EUTELSAT (line 2 of Table 8.1) is satisfactory within its service area, then an output signal power of at least $(-122 - 3) = -125$ dBW is required. Assume losses as before (column 15) of 4.5 dB. Then using Table A8.7:

dish diameter	*efficiency*	A_{eff} (dB)	*output signal power* (dBW)
180	65	+2.2	−122.3
160	65	+1.2	−123.3
120	60	−1.7	−126.2
90	55	−4.6	−129.1

These few calculations already point the way. With a high efficiency 1.8 m dish reception should be satisfactory, with a 90 cm dish, not so good. If the particular satellite is only to be used occasionally then, of course the smaller dishes may be acceptable. The chances with satellites further afield can similarly be assessed.

Finally, to gain more confidence in the assessment method, consider the master antennas for cable subscribers. These need to ensure a good picture through highly unfavourable weather conditions, even from the less powerful satellites. They are of 3 − 4 m or greater diameter. Take, for example, a 4 m dish, efficiency 60%, working to INTELSAT V F-4 which gives the UK a pfd of only −118 dBW/m^2. From Equation A8(22) and allowing only 3 dB for losses (more accurate alignment) the dish signal output is −112.2 dBW, ample for a reliable service.

8.5 Noise

The detrimental effect of electrical "noise" is mentioned in Section 2.7. In fact noise primarily determines the level of signal required from a dish. If the signal presented to a tv set is not sufficiently strong compared with the noise accompanying it, then all sorts of things can happen, from a grainy or spotty picture to one which jumps all over the screen. Accordingly, before deciding how much signal is required it is essential to assess how much noise is around.

The experts have calculated the optimum dish output level required as seen in the preceding Section. They work backwards from a knowledge of the signal-to-noise (s/n) ratio required at the input to the average tv set, calculate or estimate the noise injected at various stages, and so finally decide how strong a signal the dish should deliver. To follow this through would involve us in some complicated ways of noise assessment so it is better left alone. However, not entirely for no discussion of satellite technology would be complete without at least consideration of the types of noise and how equipment is rated. This is important to those faced with choice of equipment (e.g. LNB and receiver) for one of the important factors is noise restriction.

8.5.1 The Uninvited Guest

Electrical noise arrives from many sources, a major one being the sky itself for it generates both *atmospheric* and *galactic* noise. The first is easily appreciated for at its most powerful, lightning causes crackles on the radio or white splashes on a tv screen. At such instants the noise is greater than the wanted signal and although for a moment the programme is completely blotted out, we may not realize it because the duration is so short. Lightning can be seen but what is not so evident are the many lesser discharges occurring continuously in the atmosphere, all of which are sources of radio noise. Atmospheric noise is at its worst below about 20 MHz so is not a great problem at satellite frequencies. This is not so with galactic noise which, as one might guess, comes from the galaxy (the myriads of far-off stars). Astronomers may love this noise for it tells them much about the universe but communication engineers are not so happy with it, especially when working above 20 MHz, as we are.

Nature unfortunately has another source of noise which she rather unkindly puts in all our equipment. It is due to what is called *thermal agitation*. Think first of a large pool containing millions of minute goldfish and assume that in each tiny fraction of a second we are able to count the number within a certain distance of the water surface. Clearly, because the fish are swimming about in all directions, the number counted each time will be different, we say this is *at random*, implying that the count at any instant has no relationship with that at any other instant. The goldfish represent the electrons in a piece of electrical equipment. Electrons as shown in Section 2.8 are the minute particles which in their millions and millions make up an electric current. Hence because electrons, like the goldfish, are darting about in all directions, the number at the end surface of say, a piecé of wire, varies at random. Each electron carries a minuscule charge of electricity so the total charge (or voltage) there also varies at random. Now these rapid changes in voltage (known as *white noise*, "white" inferring that the changes occur at all frequencies) are of such tiny values that seemingly they can be ignored – but not when followed by high *amplification.* As with the signal itself we talk in terms of *noise power.*

An amplifier (Sect.4.1) is an electronic device which increases the strength or power of electrical signals up to many thousands of times and so thermal noise generated in the *input* of the amplifier itself [Ref. A8(23)] is increased in power along with the signal. The result is that the s/n ratio at the output of an amplifier is worse than that at the input.

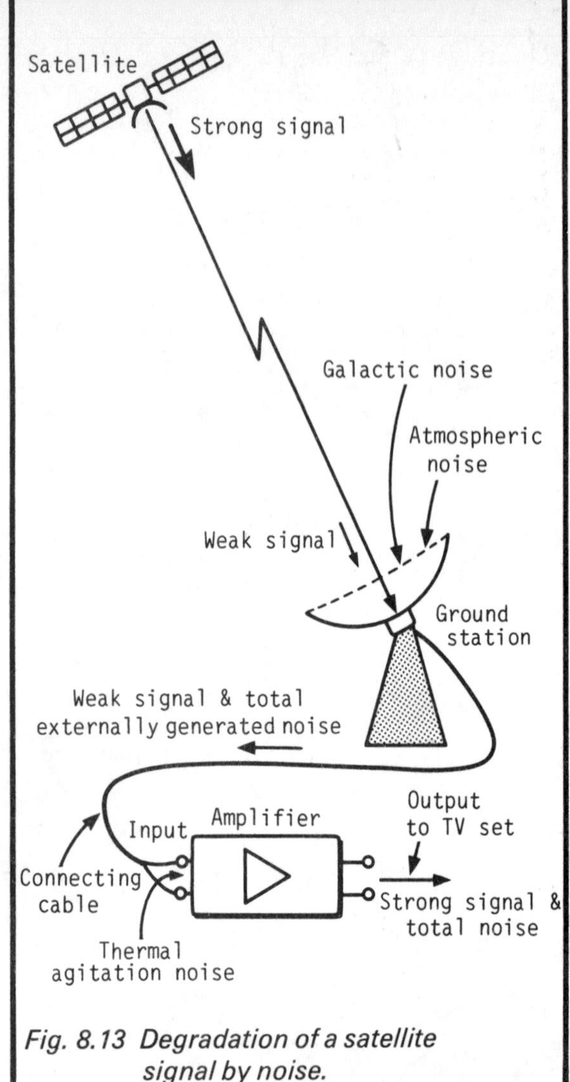

Fig. 8.13 Degradation of a satellite signal by noise.

As temperature falls the goldfish slow up and this is the same with electrons, they have less energy from heat to keep them going, hence the description "thermal" agitation. Consequently this type of noise is reduced as temperature falls and vice versa. Large Earth stations at present reduce it by working their amplifiers at extremely low temperatures but at home this is not practicable. Some amplifier components also produce noise so we rely on the manufacturers to reduce it to a minimum by their choice of components and design techniques.

Accordingly, a signal from a satellite, having gathered atmospheric and galactic noise on its way, is finally confronted by even more noise added by the receiving equipment itself. Because the total noise cannot be reduced to any great extent, arrangements must be made for the signal to be ample.

Figure 8.13 shows these considerations in outline.

8.5.2 Figures and Factors

A common way of specifying the noise performance of, for example, an LNB or receiver is by the *noise figure.* However consider *noise factor* first. This is defined as the relationship between the input and output signal-to-noise power ratios and it indicates how much the noise generated within an equipment degrades the signal-to-noise ratio as the signal passes through. Then:

$$\text{Noise factor, } F = \frac{P_{si}/P_{ni}}{P_{so}/P_{no}}$$

where P_s and P_n represent signal and noise powers and i and o refer to input and output.

For example, if the input ratio is 6 (i.e. the signal has 6 times the power of the noise) and at the output the ratio is 4, then:

$$F = \frac{6}{4} = 1.5$$

If the output ratio is reduced to 3 because greater noise is generated within the equipment, then:

$$F = \frac{6}{3} = 2$$

so the higher the value of the noise factor, the worse the equipment is from the point of view of internal noise generation.

Noise *figure* follows for it is simply noise factor expressed in decibels, i.e.

$$F_{dB} = 10 \log F$$

and the two noise factors above become noise figures of:

$$10 \log 1.5 = 1.76 \text{ dB}$$

and

$$10 \log 2 = 3.01 \text{ dB}\,.$$

Now Appendix A8.9 shows how the internally generated noise increases directly with ***absolute*** temperature, hence the temperature at which a noise figure is measured should be quoted. Take for example, two specifications for LNB's, A has a noise figure of 2.5 dB at 25°C, B's is 2.5 dB at 60°C. To compare them adjustment must be made for temperature so let us check what happens if the temperature surrounding A is raised from 25 to 60°C. The absolute or Kelvin and Celsius temperature scales have the same size degrees but the formula works in degrees Kelvin which start at absolute zero, i.e. 273 degrees below the Celsius:

$$25°\text{C} = 273 + 25 = 298 \text{ K}$$

$$60°\text{C} = 273 + 60 = 333 \text{ K}$$

(the degree symbol ° is not used with the Kelvin scale). Then increase in noise is equal to

$$\frac{333}{298} = 1.12\,,$$

i.e. about half of one dB. Accordingly the comparison is more accurately stated as:

Noise figure of A at 60°C = 3.0 dB

Noise figure of B at 60°C = 2.5 dB

and so B has the better noise performance.

Believe it or not, this is the simplest of noise evaluation methods. Communications engineers have more flexible ways of assessing system noise performance such as rating even galactic noise by an "equivalent noise temperature". This is mentioned only because it may appear in satellite literature – but it is not for us here. The technical story continues in Appendix 8 (Sect.A8.9).

Chapter 9

THE HOME TVRO

As we near the end of the book and look back it is clear that we have not only covered the basic principles but also seen how each item of equipment functions and fits in with the rest. The home installation should now seem less daunting, not child's play perhaps but certainly not beyond reach.

Before getting down to fine detail it may be profitable to take an overview of what is available. Here are the systems in order of cost:

(i) the simplest comprises a basic fixed dish, LNB and receiver for viewing certain channels on one satellite only. This is not as restrictive as it seems for with DBS all the channels of one's own country can be received in this way;

(ii) using the system of (i), other channels on the same satellite but of opposite polarization can be added. This necessitates visiting the dish outside and physically rotating the LNB;

(iii) the range of (ii) can be further increased by more visits to the dish to align it with different satellites. Although all satellites giving a sufficient signal for a particular dish can be brought in by manual adjustment of LNB and/or dish, the method is hardly to be recommended on a regular basis, lining a dish up accurately to a satellite can be a teaser. However life becomes easier if a polar mount is fitted (Sect.9.2.4). Use of the basic system therefore might be summed up as ideal for channels of the same polarization of one satellite only but less so if more is sought;

(iv) a worthwhile addition is a polarotor. This is fitted in or to the LNB and is remotely controlled from indoors to adjust the polarization, i.e. (ii) above is carried out by a flick of a switch rather than by a trip down the garden;

(v) for the enthusiast who must have it all, the most expensive systems not only have polarotors but motorized dishes as well, i.e. special equipment on the rear of the dish for swinging it to any satellite, again remotely controlled from indoors. With the more sophisticated receivers all that the user has to do is to enter the channel required. The system does the rest, it moves the dish, sets the polarity and fine-tunes everything.

Let us start outside and then work our way indoors.

9.1 Out of Doors

In Chapter 8, to get down to the nitty-gritty of collecting a share of the wave from a particular satellite, the parabolic dish antenna is examined in some detail. However, as Section 8.1.5 shows, not all dishes are parabolic, nevertheless even for these the basic transmission principles are the same. In addition dishes are manufactured in offset form (Sect.8.1.1) but with the dish profile such that it can be mounted vertically for a range of receiving locations. For these our elevation tables (A8.3 – A8.4) become redundant. Tables are also not required for the simplest of small "windowsill" dishes. These are likely to have fixed elevations pre-set for a given area. In the UK for example, working to the satellite at 31°W requires an elevation of 27.8° at Lands End and 19.1° at John O'Groat's, so several different settings are required to cover the whole country. With the elevation already set the dish is mounted on a level surface and is then rotated slowly until a picture is received on a nearby tv set. These smaller dishes have wider beamwidths and are therefore easier to align (see Fig.A6.1). Undoubtedly a dish with built-in elevation has many advantages and the system is also used on some larger ones. We continue in this Chapter however with the full procedure for lining up so that all types are covered.

What is important but has not yet been covered in detail is the way in which dishes can be realigned to other satellites. As we have seen, one can go out or aloft armed with a spanner and change to the new azimuth and elevation angles as given in the Tables[A8] but this has not been recommended for wet days, especially as polarization may also have to be changed (Sect.8.1.3). Later therefore we consider how dish manufacturers are able to supply mounts which enable a reasonably quick change to be made or for a little more expense, fit a "motorized dish actuator" so that the change is effected from the comfort of an armchair.

9.1.1 The Heavenly Arc

When or before we get up in the morning the Sun looks at us *from the horizon* in the east. At midday it is high in the sky (the meridian) and finally it disappears *at the horizon* in the west. Were we to plot its position regularly throughout the day, it would be found that the path is in the form of an arc. The Sun is in fact following a *polar curve* which is defined as one related in a particular way to a given curve and to a fixed point called a *pole* (nothing to do with the Earth's poles). Hence to follow a polar curve, not only must its shape be known but also the position of the pole. With a little imagination it is possible to see that from any earthly location the part of the geostationary orbit which contains available satellites also follows a polar curve.

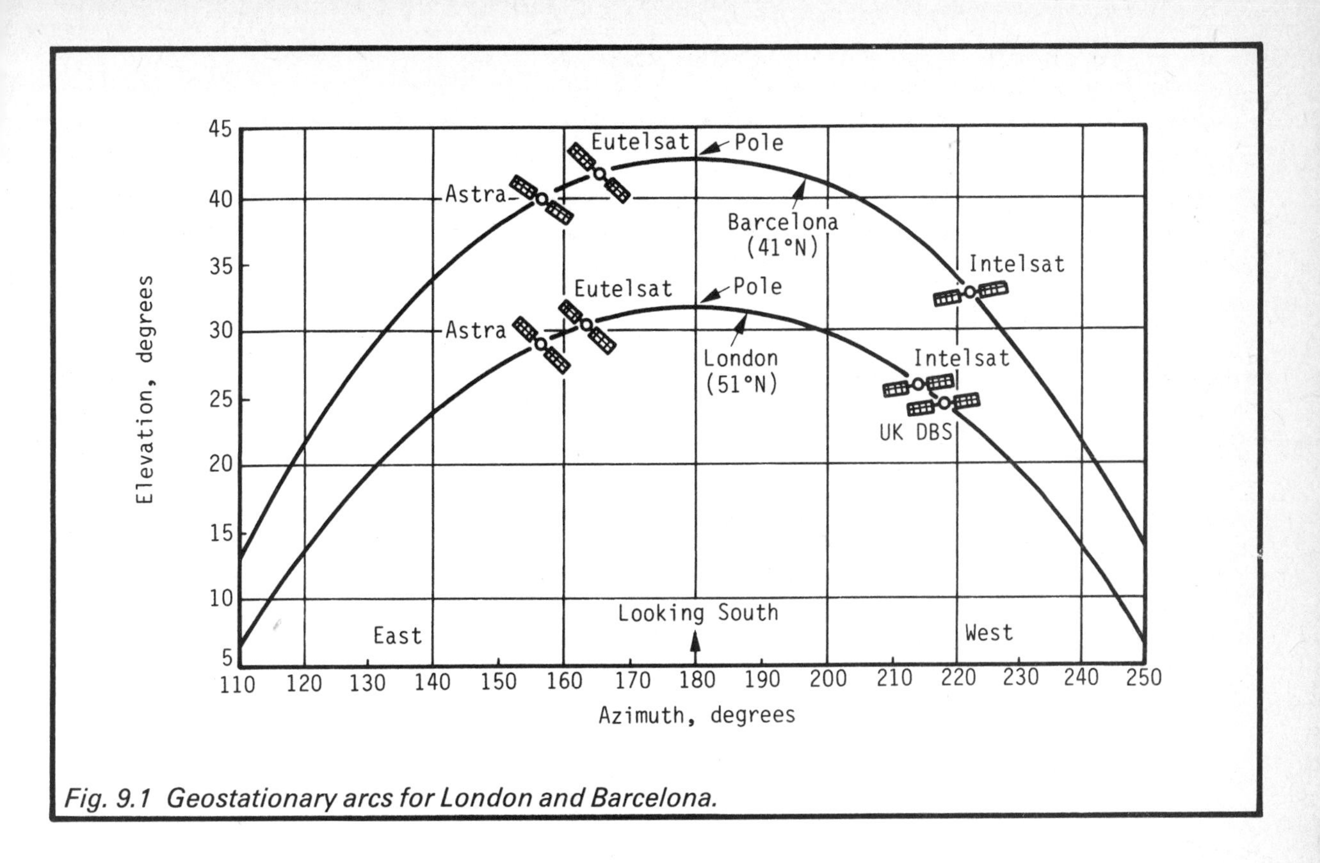

Fig. 9.1 Geostationary arcs for London and Barcelona.

Such a curve can be produced mathematically but we have an easier way out by using data from Tables A8.1 – A8.4 of Appendix 8. Take a couple of locations as an example, London at say 51° latitude, Barcelona at 41°.

Start with Table A8.1 at latitude 51°. For $\phi_d = 0$ the azimuth is 180°. A satellite in this position must be due south and at the highest point, the pole. The elevation is found from Table A8.3 and for latitude 51° and $\phi_d = 0$, this is 31.6°. Azimuth and elevation may then be plotted as for example in Figure 9.1. Plotting for other values of ϕ_d (at 51° latitude) generates the right hand half of the London curve as shown. The left part of the curve arises from Tables A8.2 and A8.4 at 51° (remember ϕ_d is negative for easterly directions). The two halves of the curve are mirror images. The curve for Barcelona is also shown in the Figure with elevations higher than those for London as would be expected. To add a little realism some satellites have been added in their present positions. An interesting exercise for it indicates the order of things pictorially. The arc is actually crowded with satellites but not all broadcast tv signals.

The arc is followed by a dish when the latter has a *polar mount*, a useful device growing in popularity. When fitted with such a mount the dish is attached to its support by a pivot inclined in such a way to the horizontal plane that when turned, a polar curve is automatically followed. The idea is not new, it has been used with telescopes for many years. Accordingly, for anywhere on Earth, rotating the dish in the horizontal plane either manually or by motor, sweeps through the arc so that the dish picks up each satellite in turn. Details on setting up a polar mount are given in Section 9.2.4.

9.1.2 Dish Choices

We ought now to recapitulate on the main types of antenna and antenna mounting available. The technical considerations of size, gain and efficiency are covered in Chapter 8 so once it is known what is required, it should be possible to choose the best compromise. But do we always know what we want? Certainly residents of a particular country will wish to receive at least the home-grown DBS transmissions. If nothing more, then a 60 or 30 cm diameter fixed dish (or the equivalent in the flat variety) will suffice and probably be the most maintenance free and least expensive of all installations. In addition, there may be a bonus. The WARC planners have grouped channels on many satellites so that it is possible to receive transmissions from adjacent countries without repositioning the antenna. We can see this from Appendix 3 where, for example the following pairs of adjacent countries, Austria and West Germany, France and Luxembourg, Switzerland and Italy, UK and Ireland have channels with the same polarization on the same satellite. When looking at footprints it must be remembered that the line of a footprint does not indicate that reception is impossible outside. The line merely indicates a certain level of power flux density which gradually reduces as the distance outwards from the line increases. Hence the fixed dish may have more to offer especially with a small increase in dish size.

Again, with a slightly larger dish, not only can other countries' DBS satellites be received but non-DBS ones as well. Altogether as time goes on, scores of channels will be available for the enthusiast. But many will be in a foreign language and others may be encrypted. So therein lies a not too easy choice, (i) the simplest of dishes perhaps installed DIY (see next Section) or (ii) something more complicated and more expensive for viewing from two, three or many more satellites. Here satellite periodicals (see Appendix 1) can help in making up one's mind for current information on programmes, languages, footprints and equipment availability is always to hand.

Given that something better than the fixed mount is required, there are again many options which generally apply to both dish and flat antennas. Basically the choice lies between:

(i) special antenna designs which increase the range over the single satellite fixed mount for example, with two LNB's fitted together in such a way that both types of polarization are received together. This system removes the need for rotating the LNB or for changing polarization by remote control. It is also possible to provide two separate feedhorns so that each "sees" a different section of the arc and can therefore pick up different satellites without having to turn the dish. Both the above methods represent an advance on the single satellite arrangement, still without the complication of dish repositioning. However they do not have the range of the more sophisticated systems;

(ii) the polar mount which can trace out the geostationary arc and is adjusted manually;

(iii) the polar mount which is operated by remote control, usually from a separate "positioner". For this an "actuator" is fitted to the mount, this usually is in the form of a motor-driven piston. A sketch of a typical actuator is given in Figure 9.2(i). As the piston is screwed in or out of the jack by the motor so the dish is turned on its pivot which is in such a position on the rear of the dish that the geostationary arc is followed. In (ii) of the Figure an idea of a positioner is given, it includes an antenna position indicator which is useful otherwise on switching on, the user has no idea as to where the dish is pointing.

If polarization changing is not catered for as in (i) and there is no desire to do the job manually then there are two choices. Either a remotely controlled motor can turn the LNB or what is more likely, a feedhorn can be used with built-in polarizer, again remotely controlled as mentioned in Section 8.1.3.

Most remote polarization control is from the receiver, usually with a "skew" control for fine tuning (Sect.8.1.4).

9.2 The DIY Approach

There is little doubt that some readers will wish to install as much of the home equipment as possible. There are several reasons, for example:

(i) it is a challenge;
(ii) it leads to a more intimate relationship with the technology;
(iii) maintenance of the outside equipment is in one's own hands;
(iv) there is also the question of cost.

Read this Section thoroughly before making a decision, calling in the dealer when nothing works might be humiliating. Here to help, is a collection of reminders, hints and tips. It cannot be a step-by-step guide because installations differ, in some cases widely. Some manufacturers supply complete DIY packages including tools, instruments and instructions. Even dish antennas can be constructed from DIY kits.

9.2.1 Site Survey

First must come a site survey unless it is evident that there is a clear wide view to the south. Section 8.1.2 covers the principles involved and to make a reliable check a compass for judging azimuth and an inclinometer for elevation are required. These need not be complicated devices, for example a pocket compass will suffice but never forget that any compass points to *magnetic* North, not the *true* North we are working with. The position of magnetic North is at present about 79°N, 70°W (NW corner of Greenland). What makes life difficult for those concerned with navigation is that it does not stay in this position but slowly wanders around a circular path of about 160 km diameter. From Figure 9.3(i) it can be seen that in the position shown the compass is pointing to west of North. In some parts of the World it may point to east of North. In the example shown the 8° difference is known as the *magnetic variation*, an apt description because it not only varies over the entire Earth but, as mentioned, also with time. It is sometimes referred to as the *magnetic deviation.*

The magnetic variation at any particular location can be obtained from the British Geological Society or from charts produced by the UK Ministry of Defence (Sect.A1.2). However, to be helpful, Table 9.1 contains the estimated variations on a latitude/longitude basis. The figures are to the nearest degree except where a .5 is used. They are estimated for January 1989 and can be expected to reduce by about 10 minutes (0.167 degrees) annually. Hence in 6 years all figures will be reduced by one. Fortunately for the whole of Table 9.1 the

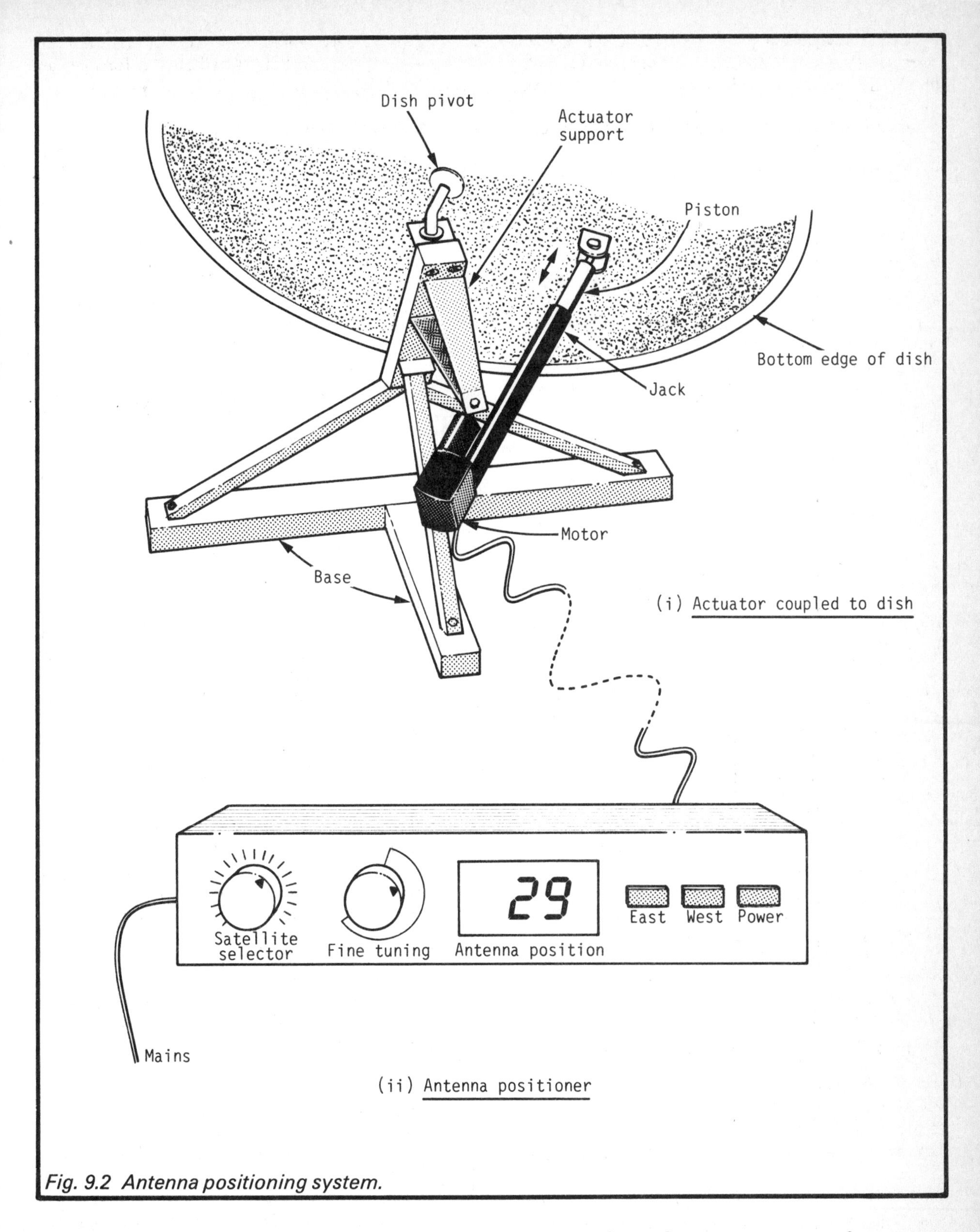

Fig. 9.2 Antenna positioning system.

variations are all westerly so we have only one rule to observe: "add westerly variation to true to find magnetic". This rule is demonstrated in Figure 9.3(i).

As an example, suppose at, say, Liverpool in the UK (53.4°N, 3.1°W) an azimuth of 210° is required. From Table 9.1 the approximate magnetic variation is 6°W, hence the compass reading should be 210° + 6° = 216° (i.e. 216° on the compass at that location indicates 210° from true North).

TABLE 9.1 ESTIMATED MAGNETIC VARIATION (JANUARY 1989)

(Figures to be added to true bearing to obtain compass bearing)

Latitude, degrees	LONGITUDE, degrees															
	← W →										← E →					
	10	9	8	7	6	5	4	3	2	1	0	1	2	3	4	5
58					9	9	8	7								
57					9	8	8	7	7							
56					8	8	7	7	6							
55		10	9	9	8	8	7	7	6	6						
54	10	10	9	9	8	7	7	6	6	6	5					
53	10	9	9	8	8	7	7	6	6	5	5	4				
52		9	9	8	7	7	6	6	6	5	5	4	4			3
51							6	6	5	5	5	4	4	3	3	3
50					7	6.5				5	4	4	4	3	3	3
49									5	5	4	4	3	3	3	2
48						6	6	5	5	4	4	4	3	3	3	
47								5	5	4	4	4	3	3	3	
46										4	4	3	3	3	2.5	
45										4	4	3	3	3	2	
44										4	4	3	3	3	2	
43	7	7	7	6	6	5	5	5	4	4	4	3	3	3	2	
42		7	7	6	6	5	5	4.5	4	4	3	3	3	3	2	
41		7	6	6	5	5	5	4	4	4	3	3	3	2		
40		7	6	6	5.5	5	5	4	4	4	3					
39		7	6	6	5	5	5	4	4	4	3					
38		7	6	6	5	5	5	4	4	4						

Although magnetic variations of this order may seem insignificant, we ignore them at our peril for 5° down here translates to only a little less than 5° up there, which could easily be a different satellite.

An inclinometer can be purchased but a wooden base, rod, spirit level and protractor, most of which are to be unearthed at home, are the sole ingredients of a home-made one, see Figure 9.3(ii). Equally a plumb line and protractor may be sufficient. Now Tables A8.1 to A8.4 are brought into use. The longitude of the satellite (ϕ_s) will be known, the latitude (θ_r) and longitude (ϕ_r) for the locality are obtained

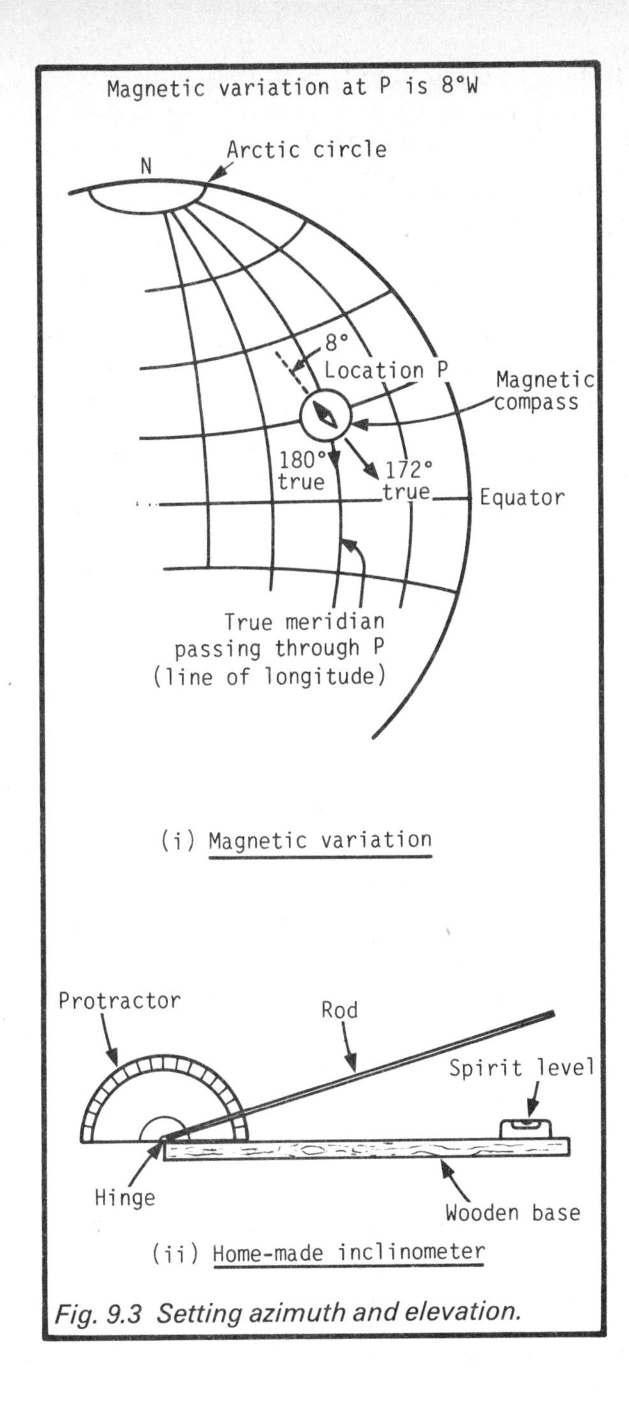

Fig. 9.3 Setting azimuth and elevation.

from a map. Longitudes east are given a negative sign. Then, as shown:A8(5)

$$\phi_d = \phi_s - \phi_r$$

from which the Tables give the azimuth and elevation angles to be used. The figures are given to one decimal place but can be rounded to the nearest degree, our instruments and techniques hardly warrant better accuracy. The inclinometer is set level and then adjusted for the elevation. Using the compass it is then rotated to the azimuth (plus magnetic variation), so then pointing approximately to the satellite. Repeat for other satellites if required.

For some of us the result of all this will be that the only antenna position feasible is on the chimney stack and for this there is obviously a limit to dish size (say 90 cm). This may bring us to a halt with our DIY intentions for the practical difficulties have now increased considerably. Not only does one have to contend with the dangers of ladders and their fixings, working at a height on anything but a flat surface, but also the added complications of lining up to the satellite. Overall, better perhaps to call in the expert installers – don't be venturesome by taking on anything which is at all dangerous.

9.2.2 Dish Installation

Once the choice of dish has been made and the goods are to hand, installation begins. We consider a simple dish for single satellite working first. It has been suggested that a pointing accuracy of within about half a degree is finally required so care has to be taken. Many different built-in aids are added by manufacturers but if there are none, we can employ the makeshift devices suggested above. After all their help is only needed in locating the satellite, accurate lining up is done afterwards on the tv picture or a signal-level meter.

If the dish mounting is a tripod or other metal base, this base must be level. If it is a circular or square tube which is sunk into the ground, this must be vertical which entails checking twice (e.g. with a builder's spirit level), the second at 90° to the first. We are assuming that the cable run from dish to house is within the generally accepted limit of 80–100 m. If greater it may be possible to overcome the excessive cable loss by fitting a special "line amplifier".

To get the azimuth setting correct it may be useful first to indicate the North–South line on the ground by pegs and fine string. Be careful not to use the compass near to the dish for any iron or steel in the structure will attract the needle and produce a wrong bearing.

Readers may find mention elsewhere of a technique of calculating when the Sun appears to be in exactly the same position in the sky as any particular satellite. As an example, for the UK DBS satellite at 31°W, this occurs at 3.00 p.m. on 15th October. At this time and provided that clouds do not obscure the Sun, then anywhere in sunshine is a suitable site for a dish, also pointing the dish towards the Sun aligns it correctly. There are two difficulties with this method:

(i) special astronomical tables are required,
(ii) expecting the Sun to shine on a certain date is, for the UK, a little optimistic.

Accordingly, we continue with the azimuth and elevation method which at least is not dependent on the weather.

9.2.3 Lining Up

To help in installation of some systems, their dealers provide installation kits on loan. Such a kit may include compass, inclinometer and even a signal strength meter which can be tuned to the appropriate satellite channel. We assume however that such help is not available.

Undoubtedly the most effective way of finalizing the dish alignment is to have the satellite receiver and a tv set connected nearby. This entails making up a coaxial cable (Sect.9.2.5) to connect the LNB to the receiver, the cable from receiver to tv set can be the one to be finally used indoors. Power has to be run out to the roof or garden for receiver and tv and great care must be taken, especially in the garden where electric shock can be lethal. On the ground common sense demands rubber or plastic boots or at least soles of these materials. Also **never** work on the live side of the mains (brown or red wire) with the power on. To cut the power off, pulling out the plug is preferable to merely switching off. It is worth mentioning that homes or sockets equipped with *residual current circuit breakers* (previously known as *earth leakage circuit breakers*) provide good electrical safety.

There is one alternative which is that a member of the family or a friend may be co-opted to watch the tv inside the house and act as a human signal level meter by calling out the state of the picture to the person struggling with the dish. No equipment is moved by this method but too much guessing has to go on for it to be recommended although, let's face it, many terrestrial tv antennas are adjusted in this way.

Ensure that the LNB is set for the appropriate polarization and offset (Sect.8.1.4), is pointing correctly to the centre of the dish and that the receiver is tuned to the desired satellite channel known to be on the air, then switch it all on. There could be a picture but don't bank on it! If only a screenful of white flashes, hold the dish from behind and make a square search of the sky until the satellite is found. This is done by moving the dish to pick out a tiny square, then on the same centre, a second one slightly larger, then again, increasing the square size and so on. A satellite must eventually be found by this method. It will first be recognized by a slight break-up of the noise flashes on the tv screen. Once this happens, further fine adjustment of azimuth, elevation, LNB and skew control should bring in the picture. Adjust for the best possible. A useful method when the picture is held is to swing the dish slightly to the left, just into noise, then over to the right again just into noise. The best picture should be about half-way between the two positions. This adjusts for azimuth. Repeat up and down for elevation. Finally tighten all adjusting nuts.

Many receivers include a signal level (or strength) meter. Using this when making adjustments for maximum signal has several advantages over watching the tv picture. A good picture may not necessarily mean that the signal is at its best and we must aim for this to have something in hand for a rainy day (Sect.8.3.2).

As an alternative, some receivers have an AGC terminal at the back. This stands for *automatic gain control.* It is an electronic system which maintains the output of a receiver fairly constant irrespective of variations in the input signal level. The signal at the output is monitored and a control voltage commensurate with its level is fed back over the agc line to the input amplifiers. Here a high level of agc voltage reduces the amplification and vice versa. The AGC terminal is connected to the agc line so connecting an ordinary voltmeter between the terminal and the ground (or earth) terminal provides a visual signal level check. The voltmeter can probably be used at some distance from the satellite receiver over a length of twin bell or telephone wire. It is worth trying out as this method may avoid connecting the whole affair up in the garden. Visits indoors or help of an accomplice will still be required to check the picture but the ultimate fine tuning is carried out more easily with the voltmeter near at hand outside.

The flat antenna should present fewer problems. Typically the plate might be only some 2 cm thick and have a simple bracket mounting at the rear. Being much lighter than an equivalent dish, it can be mounted practically anywhere where there is the mandatory uninterrupted view of the satellite(s), even on a wall or sloping roof. An added bonus is that it is said not to require such accurate alignment although it is always wise to aim for the best. Connection from the LNB to the indoor unit is the same as for a dish.

9.2.4 Polar Mounts

An introduction to the polar mount is given in Section 9.1.1. Compared with the basic *Az/El* mount, it has much to offer, not in setting up because this is more difficult but certainly in subsequent use. It has appeal to all who wish to change between satellites frequently and is a must for those wanting to do this from the armchair for it is the only form of mounting which lends itself easily to being "motorized". The shape of the polar curve has been built into the mount by the manufacturer, setting up involves aiming the dish accurately at the pole. The pole is in fact at the highest point of the geostationary arc and because the arc is symmetrical about its pole, for any location in Europe the pole must be at true South. As might be expected the elevation of the pole varies with the latitude of the dish location. Let us imagine a wanted satellite at this point.

Tables A8.3 – A8.4 can be used to find the correct elevation for a dish pointing South for then the difference in longitude, ϕ_d , between the dish location and the satellite is zero. For example, at latitude 40°N (central Spain) the elevation required is 43.7° (Table A8.3) and the dish can be set to this. However there is another method which achieves the

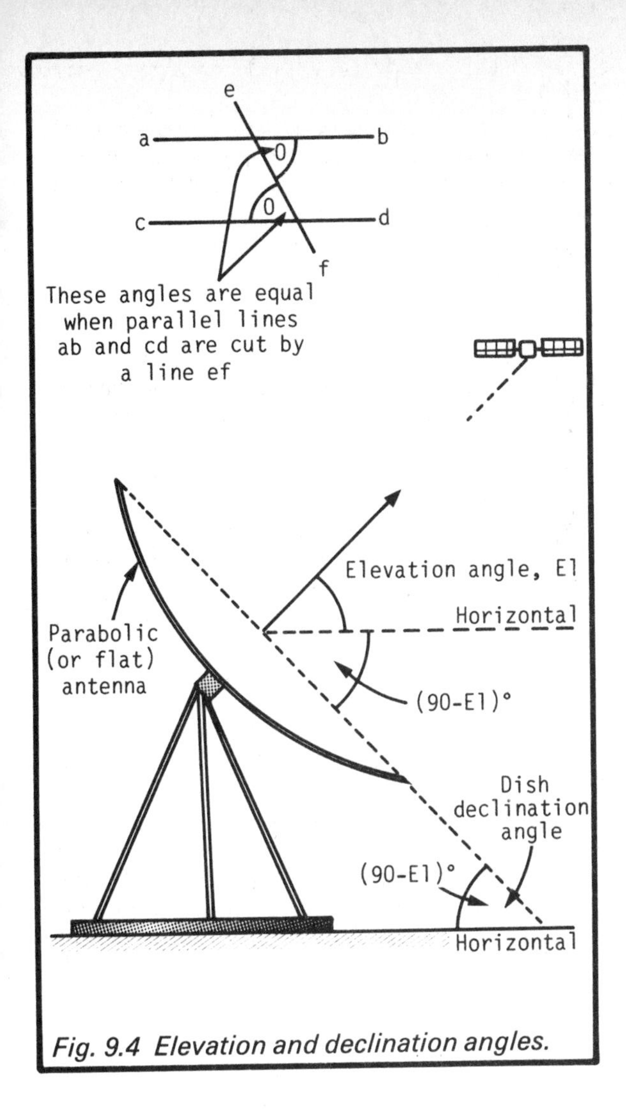

Fig. 9.4 Elevation and declination angles.

same result but with a less complicated table although it hides under a rather dissuasive title, "Dish Declination Offset Angles". The full reasoning is given in Appendix 9 (Sect.A9.1) but briefly, all that has to be done is to add the appropriate offset angle to the latitude to obtain the declination angle. Table A9.1 lists the standard offset angles for Europe. For the case quoted above (40° latitude) this one is 6.3° hence the declination angle is 40 + 6.3 = 46.3°. Most polar mounts are conveniently calibrated so that this angle can be easily set.

Figure 9.4 shows how the declination angle relates to the normal elevation angle, i.e. it is converted by subtracting it from 90° so we get 90° – 46.3° = 43.7° as above. Summing up:

latitude + offset angle (from Table A9.1)

= declination angle

elevation + declination = 90° .

A polar mount dish facing true South and so aligned, will track the whole arc from horizon to horizon as it is turned. This can be done by hand or motor. If such a dish is not accurately aligned or its support is not truly vertical then the tracking curve will not coincide with the geostationary arc, the result being a wonderfully clear view of nothing.

Although only dishes have been considered, the technique applies equally to flat antennas (Sect. 8.1.5).

Much of the basic installation work is covered in the previous two Sections but there are some differences with the polar mount. For example, as shown, elevation need not be adjusted in the same way and in fact Tables A8.1 – A8.4 are not needed. The only information required is the latitude and offset angle.

If we understand the principles involved then the installation work follows ordinary common-sense rules. Nevertheless a few additional hints may help.

Firstly the post on which the dish is mounted must be absolutely vertical. To check this a builder's spirit level is useful and the post must be checked twice, 90° apart. The polar axis of the dish must next be aligned true N–S, the stretched string pegged to two stakes in the ground may help (Sect.9.2.2). Don't forget to add in the magnetic variation (Table 9.1) when using a compass. Latitude and offset angles are easily adjusted irrespective of how the mount is marked. Elevation or declination angles can be checked by running a taut string from top to bottom of the dish and with an inclinometer measuring the angle the string makes with the horizontal (see Fig. 9.4). Remember it is quite in order to set the dish to the elevation instead of the declination angle if more convenient.

Once the dish points true South at the correct declination angle it is in fact lined up to the geostationary arc. Swinging the dish slowly round to left or right ought then to pick up a satellite near the top of the arc. When one is found, make adjustments as necessary to the mount for maximum signal as in Section 9.2.3. Then swing to the other side of the arc to pick up a second satellite and again adjust. By swinging between the two and continuing to make finer adjustments each time, the highest part of the arc should be traced satisfactorily. Then move on further to other satellites. If, when in use the polar mount is to be manually adjusted, reference marks made somewhere on the mechanism can help to ensure quick returns. When all is finished check all bolts for tightness.

Overall hardly child's play and possibly time-consuming, but rewarding. What is more, now having an in-depth and practical understanding of the whole arrangement subsequent maintenance becomes that much more manageable.

9.2.5 Cabling

All the bits and pieces have to be connected together and a length of cheap bell wire just will not do. Terrestrial tv proves this for all its antenna downleads comprise *coaxial* cables in which there are two conductors (metals along which electricity flows). One conductor is an outer tube, the other is a wire running centrally through the tube and kept in position by a special foam or solid insulating material. Such cables require special connectors. Satellite tv similarly needs coaxial cables and when the signal travels along them there is the inevitable loss (it would soon be lost altogether over a length of bell wire). Generally, for the same materials and method of construction, the larger the diameter of the cable, the lower is its loss. Typically a 10 mm diameter cable might have a loss of 12 dB per 100 metres whereas at 5 mm the loss is more than double.

Although both signal and noise suffer the same loss when travelling along a cable and on this account the signal-to-noise ratio does not change, there could be the problem of electrical noise being picked up by the cable itself. Coaxial cable has the advantage of high immunity from interfering signals and noise from outside due to its special construction. Nevertheless it cannot be completely immune hence it is important to aim for low loss and low pick-up which means installing the dish as near to the house as possible. It would be unwise therefore to keep a dish out of sight at the bottom of a long garden.

Because the fully automated dish (remote control of polarizer and actuator) needs extra wires for control, special multi-cables are available consisting of the signal coaxial cable with one or more non-coaxial cables combined.

9.3 Indoors

On moving indoors DIY ceases for all but the brave. Gone are the days of the electronics enthusiast brandishing screwdriver and soldering iron, happily wiring transistors and what goes with them to produce electronic marvels. Technology has advanced so much over the last few decades that equipment now revolves mainly round microprocessors and integrated circuits, just as computers do. Hence although construction of a satellite receiver is feasible for the owner of moderate experience, a workshop bench and a few items of test equipment (more on this in Appendix 1.2), it is not generally for DIY so the receiver normally has to be purchased *en bloc*. A typical TVRO layout is given in Figure 1.4. Here we add more detail.

9.3.1 The Satellite Receiver

Section 7.6 has already put minds at rest over the question of whether transmissions in MAC format (Sect.7.6) will make "terrestrial type" (PAL) tv sets obsolete. For the time being, adapters can be used which code the MAC signals into PAL for connection to the normal antenna socket in place of the lead from the terrestrial antenna. Alternatively the adapter circuits may be built into the receiver. Most of the benefits provided by MAC are lost, but they will be regained on replacement of the set at a later date for this will be one designed for MAC and the improved sound, probably with stereo. Installing

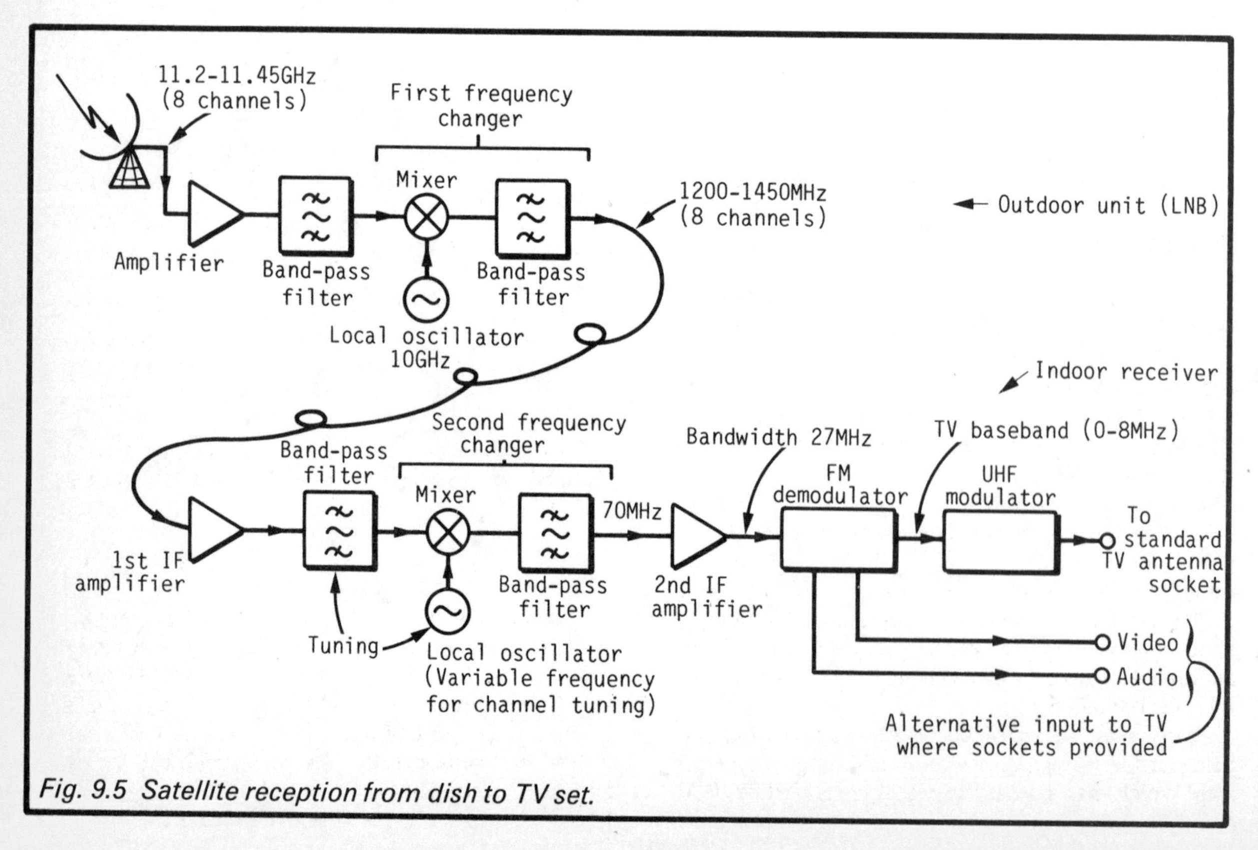

Fig. 9.5 Satellite reception from dish to TV set.

satellite tv does therefore not require change of existing tv set.

From the knowledge we have gained so far, especially with regard to signal processing (Chapter 4), it is possible to trace the progress of a satellite signal from dish to tv set. Figure 9.5 shows this typically for non-DBS reception (e.g. ASTRA, EUTELSAT, INTELSAT). The satellite signals from the feedhorn, which we imagine are for 8 separate channels, are reduced in frequency in the LNB by a frequency changer (Sect.8.1.3). We recall that carrier frequencies can be changed with the modulation unaffected – rather like shifting parcels into different vans, the van may change but the parcel does not. It therefore reaches its destination unaffected by the different vans which have carried it. The frequency changer comprises a local oscillator, mixer and following band-pass filter. The new carrier frequencies are within the range, say, 1200 to 1450 MHz and whatever channel frequencies are used, each carries the 27 MHz tv band. This band of frequencies is known as the *first intermediate frequency* (1st IF) and is suitable for transmission over the coaxial cable to the receiver indoors.

At the input of the receiver the signal is still relatively weak and so is first amplified. It then passes to a system of tuning circuits and frequency changer combined. By controlling at once the frequency of the local oscillator and the band pass filter preceding it the single channel required is selected and its carrier frequency changed to, say, 70 MHz, the 2nd IF, i.e. the channel is "tuned in". It is again amplified and then *demodulated*, the process which regains the original modulation from the final carrier of 70 MHz. The modulation frequencies cover a band containing the video and audio information for driving the tv set. Some tv sets have special input sockets for video and audio but most have only the antenna input socket. In this latter case the tv baseband signal modulates a carrier within the normal tv range which is then fed directly to the socket. The carrier frequency used is usually but not necessarily on tv channel 36 (591.25 MHz – not to be confused with satellite channel numbers). This is a channel which has no terrestrial broadcast on it and is therefore interference free. A button on the tv set must be tuned to this frequency.

Figure 9.4 shows the bare bones only, there are other features which are required for signal control, noise reduction etc. For DBS there is little change except that the receiver must match the LNB (Sect. 8.1.3) as far as the first intermediate frequency band is concerned. Also there may be need for an additional adapter for changing MAC signals to PAL for the older type tv sets.

Many may ask why if fundamental changes are being made to tv do we not take the additional step which has revolutionized disc recording (the compact disc), i.e. "go digital". Unfortunately although digital techniques have much to offer, for satellite tv each channel would need a bandwidth at present of some 100 MHz, nearly four times that required by the PAL and MAC systems (about 27 MHz). Such a bandwidth is not feasible at present.

Without doubt, satellite receivers are highly complex devices and many are microprocessor controlled. This implies that a number of interesting facilities can be provided to make the complicated business of satellite finding, polarization, polarization offset, channel tuning etc. less demanding, but at a price. Some of the facilities for a manual or fully automated system with microprocessor control are listed below:

(i) remote control of the receiver – a portable pad with buttons as is used for tv. In a way this is remote control of remote control;

(ii) the channels may be programmable; all the operations required to select each channel are entered first (e.g. antenna position, channel frequency, polarization). This data is held in an electronic memory then brought into action automatically when the required channel number is entered;

(iii) individual parameters can be entered on the receiver or remote control pad, e.g. antenna position, input frequency tuning, polarization and skew when searching for a channel not programmed as in (ii);

(iv) control of video and audio levels input to the tv set for best balance;

(v) the connexion sockets on the rear of the receiver may include one for the normal terrestrial tv antenna so that the tv set can be switched to either source without plug changing;

(vi) information showing the user that he or she is on the right track can be displayed on the front of the receiver or it can even be shown on the tv screen;

(vii) channels not suitable for the kiddies can be "locked" so that they are only accessed by entering a special code;

(viii) a signal level meter – helpful when tuning;

(ix) buttons or a rotary switch may be there so that the IF bandwidth can be changed, i.e. usually reduced. Section A8.9 in discussing electrical noise shows that the amount which gets through on a channel is proportional to the bandwidth of that channel. Accordingly if the normal bandwidth of 27 MHz is restricted, so is the noise. This facility is useful when for example we are desperate to tune in a channel from a satellite which is not among those producing a reasonable signal in the local area. Because the signal is weak, the total noise arriving with it may be sufficient to break up or otherwise spoil the picture. By switching to a reduced bandwidth, say as low as 18 MHz, the signal-to-noise ratio is improved as Section A8.9.3 shows. It is thus possible to end up with a stable picture although some of the detail will have been lost. It is a matter of compromise, we cannot have it both ways;

(x) the antenna positioner may also be built into the receiver;

(xi) the ubiquitous digital clock may also be around.

As already quoted: "You pays your money and you takes your choice".

Appendix 1

FURTHER READING AND INFORMATION

A1.1 Programme and General Information
(as at January 1989)

Satellite TV Europe: Monthly. Contains programmes for the month with general and some technical information and updating.

Cable and Satellite Express: A fortnightly newsletter, mainly for operators.

Cable and Satellite Europe Yearbook: Information covering the whole field.

World Satellite Almanac: Information about present and future satellite systems.

All the above from 21st Century Publishing, London.

What Satellite: A satellite news magazine given free with the monthly video magazine, *What Video* – WV Publications Ltd., London.

Satellite A–Z: A magazine which is a supplement to the bi-monthly magazine, *Video A–Z* – Video A–Z Ltd., London.

Broadcast: Industry weekly news magazine – International Thomson Publishing Ltd., London.

A.1.2 Technical Literature
For the technically minded, several books are published, each covering much of the whole story in its own particular way. In addition there are many technical papers on various aspects of the subject written as the technology developed. These are too numerous to be included here but references to many of them are to be found in the books quoted.

Satellite Broadcasting Systems: J.N. Slater and L.A. Trinogga (Ellis Horwood Ltd.).

Satellite Communication Systems: G. Maral and M. Bousquet (John Wiley and Sons).

Satellite Communication Systems Engineering: W. L. Pritchard and J. A. Sciulli (Prentice-Hall).

Satellite Communications: T. Pratt and C. W. Bostain (John Wiley and Sons).

Satellite Broadcasting: P. Rainger, D. Gregory, R. Harvey, A. Jennings (John Wiley and Sons).

Introducing Satellite Communications: G. B. Bleazard (NCC Publications).

Satellite TV Systems (Troubleshooting and Repairing:) Richard Maddox (TAB Books Inc.).

Indoor Unit For Satellite TV Reception: Elektor Electronics, October/November 1986, January 1987.
This is a series of articles describing the construction and operation of a satellite receiver to connect directly between the LNB and a conventional (PAL) tv receiver). Full constructional details are given together with names and addresses of component suppliers. It is reported that the unit is very successful, however it is not recommended that absolute beginners should have a go.

Elektor Electronics: Monthly – 1 Harlequin Avenue, Great West Road, Brentford.

For Information on Magnetic Variation:

The British Geological Society,
Murchison House,
West Mains Road,
Edinburgh

or British Admiralty Chart 5375 (published by the Ministry of Defence).

A1.3 Payments
In the UK a TVRO licence is required. The present once-only fee is £10 and applications are made to the Department of Trade and Industry (Radiocommunications Division).

Information regarding payments due to programme providers of Screen Sport and Lifestyle is obtainable from The Satellite Programme Clearing House, London. For Premiere and Children's Channel the offices are also in London.

Appendix 2

FORMULAE FOR CHAPTER 2

A2.1 The Earth's gravitational force of attraction (Sect.2.4.2)

$$F = \frac{G\, m_1 m_2}{d^2} \text{ newtons} \quad (1)$$

where F is the force
m_1 and m_2 are the masses of the bodies (kg)
d is the distance between them (m)
G is the gravitational constant (6.67×10^{-11} $\text{N.m}^2\text{kg}^{-2}$)

A2.2 Sidereal period of rotation of Earth (Sect.2.4.4)

$$= \frac{365.25 \times 24 \times 3600}{366.25} = 86164.1 \text{ seconds} \quad (2)$$

A2.3 Decibel notation (Sect.2.5)

Power: No. of decibels:

$$= 10 \log_{10} \frac{\text{power sent } (P_s)}{\text{power received } (P_r)} \quad (3)$$

Voltage and Current: No. of decibels:

$$= 20 \log_{10} \frac{\text{voltage or current at input}}{\text{voltage or current at output}} \quad (4)$$

(input and output measurements must be in the same impedance).

If no logarithm tables or suitable calculator or computer are available, Table A2.1 can be used to determine the approximate power ratio for any dB value. Some examples are:

(i) 17 dB = 10 dB + 7 dB

converted to power ratio = 10 x 5.01 = 50.1

(ii) 35 dB = 30 dB + 5 dB

converted to power ratio
= 1000 x 3.16 = 3160

(iii) 206 dB = 200 dB + 6 dB

converted to power ratio
= $10^{20} \times 3.98$

this can remain in scientific notation as 3.98×10^{20}

(iv) –59 dB = –60 dB + 1 dB

converted to power ratio
= $10^{-6} \times 1.259 = 1.259 \times 10^{-6}$

TABLE A2.1
APPROXIMATE RELATIONSHIP BETWEEN DECIBELS AND POWER RATIOS

dB	Power ratio P_s/P_r	dB	Power ratio P_s/P_r
–10	0.1	5	3.16
– 5	0.32	6	3.98
– 3	0.5	7	5.01
0	1.0	8	6.31
0.2	1.05	9	7.94
0.5	1.12	10	10
1.0	1.26	20	100
2	1.58	30	1000
3	2.0	40	10^4
4	2.51		↓

thereafter for multiples of 10 dB the exponent of 10 in the ratio is the same as the tens figure of the dB value, e.g. 200 dB = 10^{20}, –200 dB = 10^{-20} (see Sect.2.1 for scientific notation)

Appendix 3

EUROPEAN DBS SATELLITES
(Proposals by World Administrative Radio Conference, 1977)

There are four satellite orbital positions, 37°W, 31°W, 19°W, 5°E. The list below is in country alphabetical order. For channel frequencies see Section 3.4.

COUNTRY	SATELLITE POSITION	CHANNELS	POLARIZATION (all circular)
Andorra	37°W	4, 8, 12, 16, 20	LH
Austria	19°W	4, 8, 12, 16, 20	LH
Belgium	19°W	21, 25, 29, 33, 37	RH
Cyprus	5°E	21, 25, 29, 33, 37	RH
Denmark	5°E	12, 16, 20, 24, 27, 35, 36	LH
Finland	5°E	2, 6, 10, 22, 26	LH
France	19°W	1, 5, 9, 13, 17	RH
Greece	5°E	3, 7, 11, 15, 19	RH
Iceland	31°W 5°E	21, 25, 29, 33, 37 23, 31, 39	LH RH
Ireland	31°W	2, 6, 10, 14, 18	RH
Italy	19°W	24, 28, 32, 36, 40	LH
Liechtenstein	37°W	3, 7, 11, 15, 19	RH
Luxembourg	19°W	3, 7, 11, 15, 19	RH
Monaco	37°W	21, 25, 29, 33, 37	RH
Netherlands	19°W	23, 27, 31, 35, 39	RH
Norway	5°E	14, 18, 28, 32, 38	LH
Portugal	31°W	3, 7, 11, 15, 19	LH
San Marino	37°W	1, 5, 9, 13, 17	RH
Spain	31°W	23, 27, 31, 35, 39	LH
Sweden	5°E	4, 8, 30, 34, 40	LH
Switzerland	19°W	22, 26, 30, 34, 38	LH
Turkey	5°E	1, 5, 9, 13, 17	RH
United Kingdom	31°W	4, 8, 12, 16, 20	RH
Vatican	37°W	23, 27, 31, 35, 39	RH
West Germany	19°W	2, 6, 10, 14, 18	LH

(The USSR has an allocation of 65 channels spread over satellites at orbital positions 23, 44, 74, 110 and 140 degrees East.)

Appendix 4

FORMULAE FOR CHAPTER 4

A4.1 Frequency Modulation (Sect.4.4)
Consider a carrier wave of frequency, f_c modulated by a sinusoidal wave of lower frequency, f_m with the equation, $v = V_m \sin \omega_m t$ where $\omega_m = 2\pi f_m$. Let the frequency deviation $= \Delta f$.

Since Δf varies according to the amplitude of the modulating wave, the deviation at any instant is $\Delta f \sin \omega_m t$

$\therefore$ Instantaneous frequency of fm wave (f) = (nominal frequency + deviation), i.e.

$$f = f_c + \Delta f \sin \omega_m t \qquad (1)$$

which gives the modulated wave frequency at any time, t.

Determining the *components* of an fm wave is a little complicated mathematically because the wave does not have a constant angular velocity. The outcome of doing so however is as follows.

Components of fm wave where V_c is the carrier amplitude and m the modulation index:

$$m = \frac{\Delta f}{f_m} \qquad (2)$$

(i) the carrier at an amplitude $V_c\left(1 - \frac{m^2}{4}\right)$

(ii) a pair of side frequencies $(f_c \pm f_m)$, amplitude $mV_c/2$

(iii) a second pair of side frequencies $(f_c \pm 2f_m)$, amplitude $m^2 V_c/8$

(iv) higher order side frequencies, theoretically to infinity, $(f_c \pm 3f_m)$, $(f_c \pm 4f_m)$,

etc. but reducing rapidly in amplitude and therefore can usually be neglected. This is a pointer to the fact that fm requires a large bandwidth which is generally of the order of 2 (peak frequency deviation + modulating frequency), i.e.

$$\text{bandwidth} \approx 2(\Delta f_{max} + f_m) \text{ Hz} \qquad (3)$$

As mentioned in the main text, the maximum frequency deviation is set in the design stage. For an fm radio broadcast transmission a value of 75 kH is frequently quoted. If the maximum modulating frequency is 15 kHz then

$$\text{bandwidth} \approx 2(\Delta f_{max} + f_m) = 2(75 + 15)$$

$$= 180 \text{ kHz}.$$

For tv a much greater value of Δf_{max} is required, say 6.5 MHz (13 MHz peak-peak) then for a maximum modulating frequency of, say 7 MHz

$$\text{bandwidth} \approx 2(6.5 + 7) = 27 \text{ MHz}.$$

(Note that this is a theoretical consideration only. In practice other values of deviation may be used with the bandwidth restricted to 27 MHz. The actual values for any particular system are a compromise between signal power, overall quality, noise and bandwidth, hence Equation (3) can only be taken as a guide.)

Appendix 5

FORMULAE FOR CHAPTER 5

A5.1 Motion in a Circle (Sect.5.1.1)
Consider a body moving around the circumference of a circle of radius r. The angular velocity ω is equal to

$$\frac{\text{angle moved (radians)}}{\text{time taken (seconds)}} \text{ rads/s}$$

from which it follows that $v = \omega r$ where v is the velocity of the body. This is not complete because velocity is a vector quantity. In this case the body is continually changing direction from a straight line path to a curved one. Changing velocity implies acceleration and it can be shown that there is a continual acceleration towards the centre of the circle of

$$\omega^2 r \quad \text{or} \quad \frac{v^2}{r}\ .$$

Therefore, from the basic formula

$$\text{force } (F) = \text{mass } (m) \times \text{acceleration } (a)$$

$$F = \frac{mv^2}{r} \text{ newtons } (m \text{ in kg, } r \text{ in metres}) \qquad (1)$$

This force F is the centripetal force required to restrain the body from releasing itself from circular motion and flying off at a tangent.

A5.2 Conditions for the Geostationary Orbit
From Equation (1), centripetal force required

$$F_c = \frac{m_s v^2}{r_0}$$

where m_s is the mass of the satellite (kg) and r_0 is the radius of the orbit (m), and from A2(1) the Earth's gravitational force of attraction

$$F_g = \frac{G m_s m_e}{r_0^2}$$

where m_e is equal to the mass of the Earth (kg) and G is the gravitational constant. These two forces must be equal, hence:

$$\frac{m_s v^2}{r_0} = \frac{G.m_s m_e}{r_0^2}$$

$$\therefore \quad v = \sqrt{\frac{G.m_e}{r_0}} \qquad (2)$$

The orbital period,

$$T = \frac{2\pi r_0}{v} = 86164 \text{ secs (Sect.A2.2)}$$

$$\therefore \quad T = 2\pi r_0 \times \sqrt{\frac{r_0}{G.m_e}}$$

$$\therefore \quad T^2 = 4\pi^2 r_0^2 \times \frac{r_0}{G.m_e} = \frac{4\pi^2 r_0^3}{G.m_e}$$

from which

$$r_0 = \sqrt[3]{\frac{G.m_e\,.\,T^2}{4\pi^2}} \qquad (3)$$

From Section 2.4, $m_e = 5.974 \times 10^{24}$ kg, also

$$G = 6.672 \times 10^{-11}$$

$$\therefore \quad r_0 = \sqrt[3]{\frac{6.672 \times 10^{-11} \times 5.974 \times 10^{24} \times 86164^2}{4\pi^2}}$$

$$= 4.21636 \times 10^7 \text{ m} = 42163.6 \text{ km}$$

Now r_0 extends from the centre of the Earth to the orbit, let $r_0 = (r + h)$ where r is the radius of the Earth at the equator (6378 km – Sect.2.4.1) and h is the height of the satellite.
Then

$$h = r_0 - r = 42163.6 - 6378 = \underline{35{,}786 \text{ km}}$$

give or take a km or two. In practice G is slightly modified because of the non-spherical shape of the Earth. Then:

$$v = \frac{2\pi r_0}{T} = \frac{2\pi \times 42163.6 \times 3600}{86164}$$

$$= \underline{11{,}069 \text{ km/hour}}\ .$$

A5.3 Rocket Thrust and Velocity (Sect.5.2)
Thrust developed by rocket motor:

$$F = V_e \frac{df}{dt} \text{ newtons} \qquad (4)$$

where V_e = exhaust gas velocity (m/s) and df/dt = rate of propellant consumption (kg/s).

Rocket velocity in terms of amount of propellant used:

$$v = V_e \log M \tag{5}$$

where M is the *mass ratio* which is equal to

$$\frac{\text{rocket mass before use of propellant}}{\text{rocket mass after use of propellant}}$$

Exhaust gas velocity:
Equating the kinetic energy as a function of velocity, with that for a monatomic gas molecule as a function of temperature:

$$\frac{1}{2} m V_e^2 = \frac{3}{2} kT$$

where T = temperature of burn in degrees Kelvin
m = molecular mass of gas molecules
k = Boltzmann's constant (1.38×10^{-23} J/K)

then

$$V_e = \sqrt{\frac{3kT}{m}} \tag{6}$$

A5.4 The Elliptical Orbit (Sect.5.3)

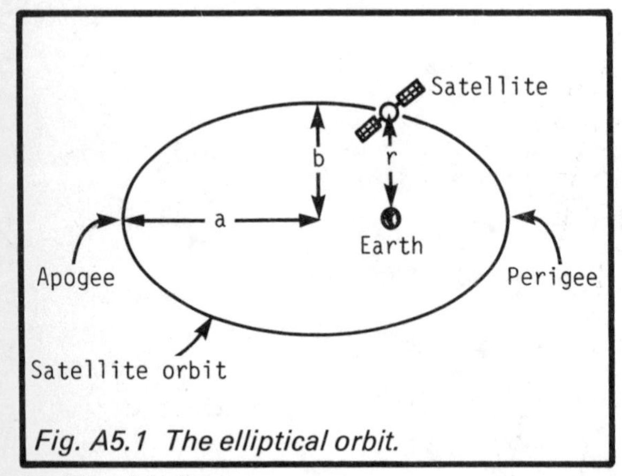

Fig. A5.1 The elliptical orbit.

See Fig.A5.1 above:
The general equation to the ellipse is:

$$\frac{x^2}{a^2} + \frac{y^2}{b^2} = 1 \tag{7}$$

when $a = b$ this reduces to:

$$x^2 + y^2 = a^2 \tag{8}$$

which is the equation to the circle.

Satellite velocity in an elliptical orbit with Earth at one focus:

$$v = \sqrt{G.m_e\left(\frac{2}{r} - \frac{1}{a}\right)} \tag{9}$$

where G is the gravitational constant and m_e is equal to the mass of the Earth.

A5.5 Surface Velocity of Earth (Sect.5.3)

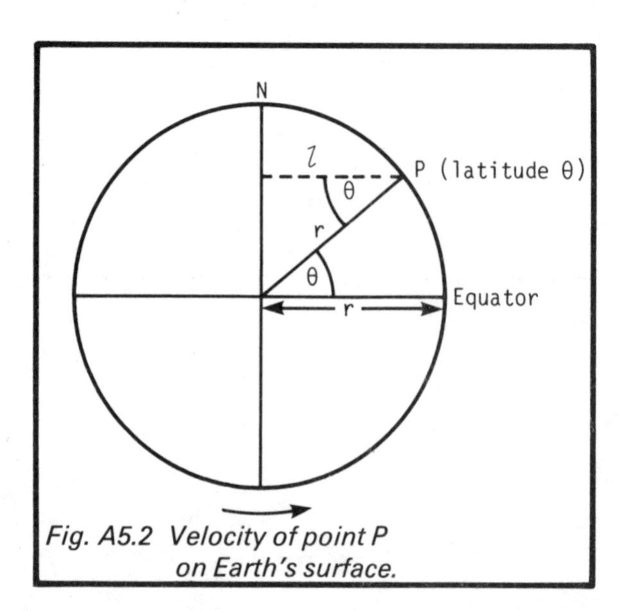

Fig. A5.2 Velocity of point P on Earth's surface.

From Figure A5.2, $l = r \cos\theta$, and distance P travels in one revolution $= r \cos\theta \times 2\pi$

$$\therefore \text{ velocity of } P = \frac{2\pi \times 6378 \times 10^3 \cos\theta}{86164.1} \text{ m/s}$$

$$= \underline{465 \cos\theta \text{ m/s}} \tag{10}$$

(Equation A2(2) and Sect.2.4) where θ is the latitude of the location under consideration.

Appendix 6

FORMULAE FOR CHAPTER 6

A6.1 Parabolic Antenna Beamwidth (Sect.6.4.1)
The theoretical width of the radiation beam of a transmitting parabolic antenna can be calculated. In this case however the formula is developed from a reasoning which involves Bessel functions, hence we must decline to get too deeply involved. At least we can use the formula to produce graphs of the beams of two practical antennas so that the effect of antenna size can be judged. Imagining the beam to be in the shape of a cone, let β degrees be half the cone apex angle, i.e. the angle with respect to the principal axis of the antenna aperture (see Fig.8.2). The signal strengths at the edges of the cone can then be compared by:

$$E_\beta = J_1\left[\frac{\pi d \sin\beta}{\lambda}\right] \times \frac{2\lambda \operatorname{cosec}\beta}{\pi d} \qquad (1)$$

where E_β is the signal strength relative to the maximum, β is the angle in degrees with respect to the principal axis of the antenna aperture, d is the antenna diameter (cm), λ is the transmission wavelength (cm) and J_1 is the first order Bessel function.

Note: the expression within the square brackets must first be evaluated, the Bessel function subsequently being obtained either mathematically or from published tables. Note also that when comparing two dishes of different sizes their maximum signal strengths are different, this is omitted from this calculation because we are only concerned with the effect of the angle β.

Consider two dishes, 60 cm and 120 cm diameter, both at 12 GHz for which $\lambda = 2.5$ cm. A typical calculation is as follows:

120 cm dish for $\beta = 0.6°$ (i.e. cone apex angle = 1.2°)

$$J_1\left[\frac{\pi d \sin\beta}{\lambda}\right] = J_1\left[\frac{\pi \times 120 \times \sin 0.6°}{2.5}\right]$$

$$= J_1[1.579]\ .$$

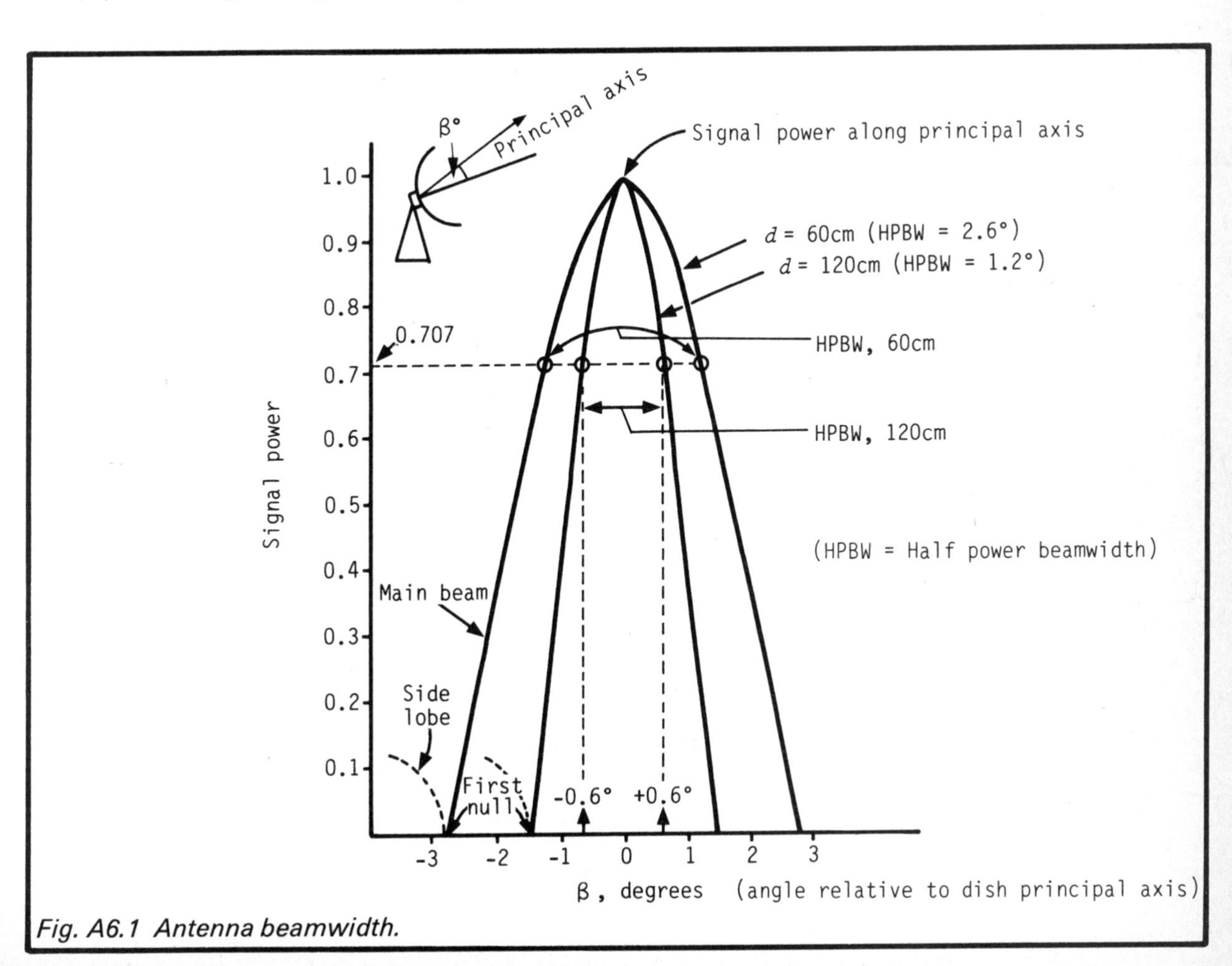

Fig. A6.1 Antenna beamwidth.

From Bessel Function Tables for J_1 we get 0.5595.

$$\therefore \quad E_\beta = 0.5595 \times \frac{2 \times 2.5 \times \operatorname{cosec} 0.6^\circ}{\pi \times 120}$$

$$= 0.5595 \times 1.267 = 0.71$$

Repeating this as required enables us to draw the two graphs shown in Figure A6.1. The effect of using an antenna of larger diameter is shown clearly in that its beam concentration is greater. By continuing the calculations beyond the value of β where the transmitted signal is zero (the first *null*), small *side lobes* appear on the graph but their levels are considerably lower than that of the main beam.

If we now consider the first null, i.e. the angle outside of which there is very little radiation, this must occur for any antenna when the Bessel function becomes zero, at which $J_1 = 3.83$, then

$$\frac{\pi d}{\lambda} \times \sin \beta = 3.83$$

$$\therefore \quad \beta = \sin^{-1}\left(\frac{3.83 \times \lambda}{\pi d}\right) = \sin^{-1}\left(1.22 \times \frac{\lambda}{d}\right)$$

i.e. β increases as $\frac{\lambda}{d}$ increases,

hence for a narrow beamwidth, $\frac{d}{\lambda}$ should be high.

These conclusions apply equally to receiving antennas for which it is important to receive as little radiation as possible off the principal axis.

Appendix 7

COMMUNICATIONS AND TV ORGANIZATIONS

This is a list of some of the major organizations concerned with communications and satellites:

WARC (World Administrative Radio Conference) decides on the radio regulations controlling the allocation of frequencies to the various types of user.

ITU (International Telecommunications Union) is the governing organization for public telephone systems throughout the world.

INTELSAT (International Telecommunications Satellite Organization) has more than 100 members and is responsible for the design and operation of systems for its members.

EUTELSAT (European International Satellite Organization) has more than 20 members and provides telecommunication services by satellite including the exchange of tv programmes.

EBU (European Broadcasting Union), the body comprising the national broadcasters of Europe.

NASA (National Aeronautics and Space Administration), the well known US Government agency dealing with space exploration.

ESA (European Space Agency), is concerned with space research, communications and broadcast satellites.

SES (Société Européenne des Satellites), the organization headquartered in Luxembourg and responsible for the Ariane series of satellite launchers. One of the many satellites being placed in orbit is ASTRA.

Appendix 8

FORMULAE FOR CHAPTER 8

A8.1 The Parabola (Sect.8.1.1)
The profile of a parabola is such that for any point P on it, the distance from a fixed point on the x axis is equal to the perpendicular distance from a fixed straight line parallel to the y axis (the *directrix*). In Figure A8.1 for example, $Pf = PD$. The fixed point on the axis is known as the focus. The equation to the curve is:

$$y^2 = 4fx \quad (1)$$

and that in the Figure is plotted for $f = 20$. From the formula it is evident that:

(i) for every +ve value of x, y has two values (+ and –)

(ii) f is always on the concave side of the curve because for –ve values of x, y is imaginary.

The angle θ is given from the value of x or y:

$$\theta = 2 \tan^{-1} \sqrt{\frac{x}{f}} \quad (2)$$

$$\theta = 2 \tan^{-1} \frac{y}{2f} \quad (3)$$

Also the distance of P from the focus,

$$Pf = f \sec^2 (\theta/2) \quad (4)$$

A8.2 Azimuth and Elevation (Sect.8.1.2)

Latitude of receiving station $= \theta_r$, longitude, ϕ_r

Longitude of satellite $= \phi_s$ (East is –ve).

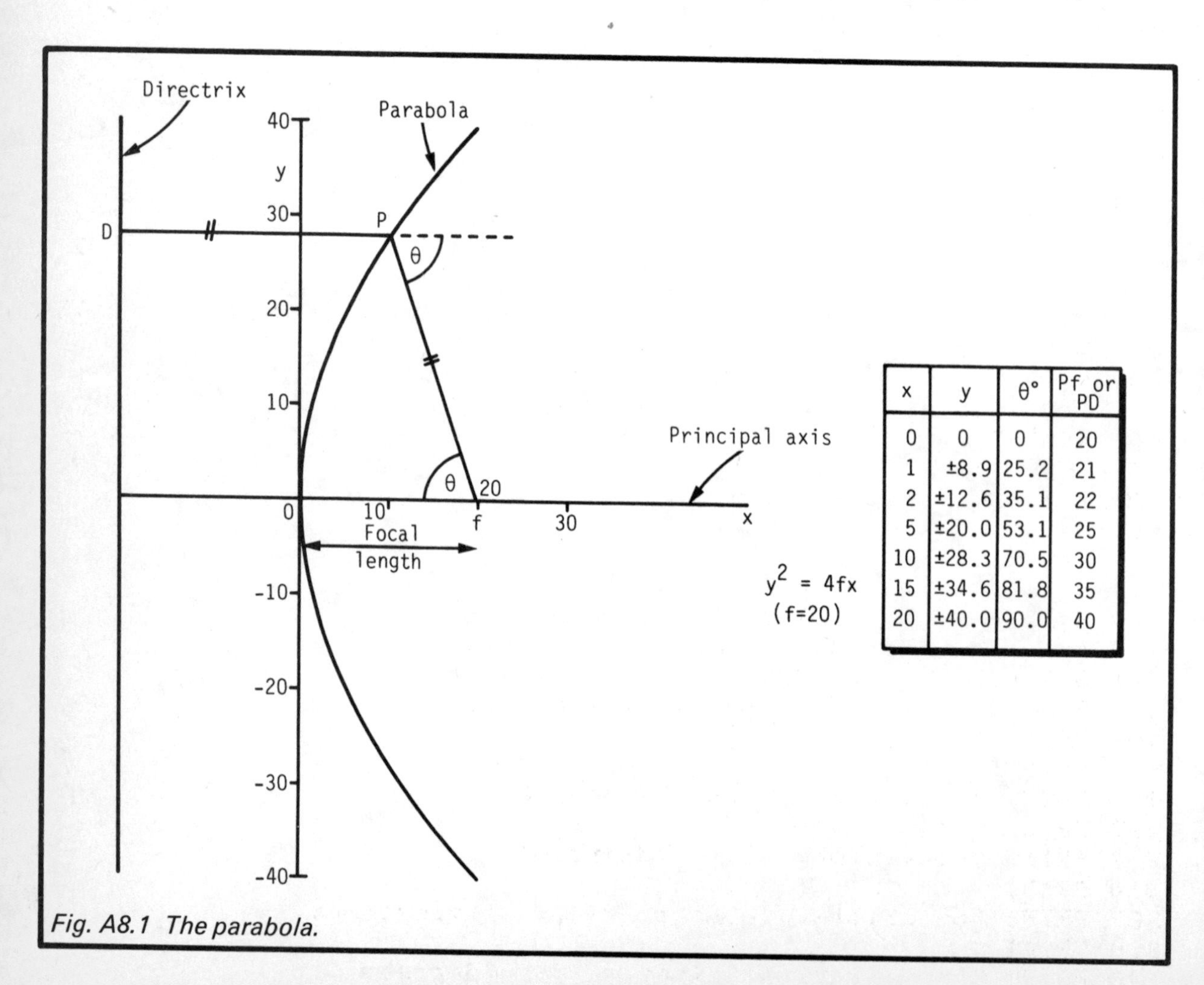

x	y	θ°	Pf or PD
0	0	0	20
1	±8.9	25.2	21
2	±12.6	35.1	22
5	±20.0	53.1	25
10	±28.3	70.5	30
15	±34.6	81.8	35
20	±40.0	90.0	40

Fig. A8.1 The parabola.

Longitude difference,

$$\phi_d = \phi_s - \phi_r \qquad (5)$$

Let r = radius of Earth (6378 km – Sect.2.4.1) and h = height of satellite (35786 km – Sect.5.1.2).

Let $x = \cos^{-1}(\cos\theta_r \,.\, \cos\phi_d)$, then Azimuth angle,

$$Az = \tan^{-1}\left(\frac{\tan\phi_d}{\sin\theta_r}\right) + 180 \text{ degrees} \qquad (6)$$

Elevation angle,

$$El = \tan^{-1}\left(\cot x - \frac{r}{r+h} \,.\, \operatorname{cosec} x\right)$$

$$= \tan^{-1}(\cot x - 0.1513 \operatorname{cosec} x)$$

or

$$\tan^{-1}\left(\frac{1}{\tan x} - \frac{0.1513}{\sin x}\right) \qquad (7)$$

($\cos^{-1}$ and $\tan^{-1}$ are synonymous with arccos and arctan).

From these formulae, Tables A8.1 to A8.4 have been calculated and examples of their use are given in the main text (Sect.8.1.2).

For readers with BASIC computers the following program is suggested. BASIC contains no Greek letters so the following variable names are used:

for θ_r, *lat*
for ϕ_r, *long*
for ϕ_s, *longsat*

The program may need modifications to suit some micro-computers but it is unlikely that the main calculations (lines 100–150) will be affected – in this area most BASIC's agree. Because many do not include *arccos* and also work in radians rather than degrees, the programmed calculations may appear to be more complicated than they really are.

It will be noticed that the program also calculates satellite distance and polarization offset. These are required for Sections 8.3.2 and 8.1.4 and are included so that one program can produce all the information needed.

A typical run of the program is shown below. The figures, e.g. 51.3, 0.1 and −13 are entered by the user when the request appears on the screen.

run
latitude of receiver 51.3
longitude of receiver (East is negative) 0.1
longitude of satellite (East is negative) . . . . . −13

Azimuth	= 163.4 degrees
Elevation	= 30 degrees
Satellite distance	= 38616.2 kilometres
Polarization Offset	= −10.3 degrees

```
10 REM SATFIND
20 REM CALCULATION OF EL, AZ, SATELLITE DISTANCE, POLARIZ. OFFSET
30 REM All entries in degrees (with decimal fractions)
40 INPUT "latitude of receiver.....",lat
50 INPUT "longitude of receiver (East is negative).....",long
60 INPUT "longitude of satellite (East is negative).....",longsat
70 pi=3.14159
80 ld=(longsat-long)*pi/180
90 la=lat*pi/180
100 x=COS(la)*COS(ld)
110 z=-ATN(x/SQR(-x*x+1))+pi/2
120 E=ROUND(ATN(1/TAN(z)-0.1513/SIN(z))*180/pi,1)
130 A=ROUND((ATN(TAN(ld)/SIN(la))+pi)*180/pi,1)
140 D=ROUND(35789!*SQR(1+0.42*(1-COS(la)*COS(ld))),1)
150 P=ROUND(ATN(SIN(ld)/TAN(la))*180/pi,1)
160 PRINT
170 PRINT
180 PRINT "Azimuth    = ",A  "degrees"
190 PRINT
200 PRINT "Elevation = ",E  "degrees"
210 PRINT
220 PRINT "Satellite distance  = ",D  "kilometres"
230 PRINT
240 PRINT "Polarization Offset = ",P  "degrees
```

TABLE A8.1 AZIMUTH (ϕ_d POSITIVE)

Latitude, θ_r, degrees	LONGITUDE DIFFERENCE, ϕ_d, degrees														
	0	5	10	15	20	25	30	35	40	45	50	55	60	65	70
38	180	188.1	196.0	203.5	210.6	217.1	223.2	228.7	233.7	238.4	242.7	246.7	250.4	254.0	257.4
39	180	187.9	195.7	203.1	210.0	216.5	222.5	228.1	233.1	237.8	242.2	246.2	250.0	253.6	257.1
40	180	187.8	195.3	202.6	209.5	216.0	221.9	227.4	232.5	237.3	241.7	245.8	249.6	253.3	256.8
41	180	187.6	195.0	202.2	209.0	215.4	221.3	226.9	232.0	236.7	241.2	245.3	249.3	253.0	256.6
42	180	187.4	194.8	201.8	208.5	214.9	220.8	226.3	231.4	236.2	240.7	244.9	248.9	252.7	256.3
43	180	187.3	194.5	201.4	208.1	214.4	220.2	225.8	230.9	235.7	240.2	244.5	248.5	252.4	256.1
44	180	187.2	194.2	201.1	207.7	213.9	219.7	225.2	230.4	235.2	239.8	244.1	248.1	252.1	255.8
45	180	187.1	194.0	200.8	207.2	213.4	219.2	224.7	229.9	234.7	239.3	243.7	247.8	251.8	255.6
46	180	186.9	193.8	200.4	206.8	213.0	218.8	224.2	229.4	234.3	238.9	243.3	247.4	251.5	255.3
47	180	186.8	193.6	200.1	206.5	212.5	218.3	223.8	228.9	233.8	238.5	242.9	247.1	251.2	255.1
48	180	186.7	193.3	199.8	206.1	212.1	217.8	223.3	228.5	233.4	238.1	242.5	246.8	250.9	254.9
49	180	186.6	193.2	199.5	205.7	211.7	217.4	222.9	228.0	233.0	237.7	242.1	246.5	250.6	254.6
50	180	186.5	193.0	199.3	205.4	211.3	217.0	222.4	227.6	232.5	237.3	241.8	246.1	250.3	254.4
51	180	186.4	192.8	199.0	205.1	211.0	216.6	222.0	227.2	232.1	236.9	241.4	245.8	250.1	254.2
52	180	186.3	192.6	198.8	204.8	210.6	216.2	221.6	226.8	231.8	236.5	241.1	245.5	249.8	254.0
53	180	186.3	192.5	198.5	204.5	210.3	215.9	221.2	226.4	231.4	236.2	240.8	245.2	249.6	253.8
54	180	186.2	192.3	198.3	204.2	210.0	215.5	220.9	226.0	231.0	235.8	240.5	245.0	249.3	253.6
55	180	186.1	192.1	198.1	204.0	209.7	215.2	220.5	225.7	230.7	235.5	240.2	244.7	249.1	253.4
56	180	186.0	192.0	197.9	203.7	209.4	214.9	220.2	225.3	230.3	235.2	239.9	244.4	248.9	253.2
57	180	186.0	191.9	197.7	203.5	209.1	214.5	219.9	225.0	230.0	234.9	239.6	244.2	248.6	253.0
58	180	185.9	191.7	197.5	203.2	208.8	214.2	219.5	224.7	229.7	234.6	239.3	243.9	248.4	252.8

TABLE A8.2 AZIMUTH (ϕ_d NEGATIVE)

Latitude, θ_r, degrees	LONGITUDE DIFFERENCE, ϕ_d, degrees														
	−0	−5	−10	−15	−20	−25	−30	−35	−40	−45	−50	−55	−60	−65	−70
38	180	171.9	164.0	156.5	149.4	142.9	136.8	131.3	126.3	121.6	117.3	113.3	109.6	106.0	102.6
39	180	172.1	164.3	156.9	150.0	143.5	137.5	131.9	126.9	122.2	117.8	113.8	110.0	106.4	102.9
40	180	172.2	164.7	157.4	150.5	144.0	138.1	132.6	127.5	122.7	118.3	114.2	110.4	106.7	103.2
41	180	172.4	165.0	157.8	151.0	144.6	138.7	133.1	128.0	123.3	118.8	114.7	110.7	107.0	103.4
42	180	172.6	165.2	158.2	151.5	145.1	139.2	133.7	128.6	123.8	119.3	115.1	111.1	107.3	103.7
43	180	172.7	165.5	158.6	151.9	145.6	139.8	134.2	129.1	124.3	119.8	115.5	111.5	107.6	103.9
44	180	172.8	165.8	158.9	152.3	146.1	140.3	134.8	129.6	124.8	120.2	115.9	111.9	107.9	104.2
45	180	172.9	166.0	159.2	152.8	146.6	140.8	135.3	130.1	125.3	120.7	116.3	112.2	108.2	104.4
46	180	173.1	166.2	159.6	153.2	147.0	141.2	135.8	130.6	125.7	121.1	116.7	112.6	108.5	104.7
47	180	173.2	166.4	159.9	153.5	147.5	141.7	136.2	131.1	126.2	121.5	117.1	112.9	108.8	104.9
48	180	173.3	166.7	160.2	153.9	147.9	142.2	136.7	131.5	126.6	121.9	117.5	113.2	109.1	105.1
49	180	173.4	166.8	160.5	154.3	148.3	142.6	137.1	132.0	127.0	122.3	117.9	113.5	109.4	105.4
50	180	173.5	167.0	160.7	154.6	148.7	143.0	137.6	132.4	127.5	122.7	118.2	113.9	109.7	105.6
51	180	173.6	167.2	161.0	154.9	149.0	143.4	138.0	132.8	127.9	123.1	118.6	114.2	109.9	105.8
52	180	173.7	167.4	161.2	155.2	149.4	143.8	138.4	133.2	128.2	123.5	118.9	114.5	110.2	106.0
53	180	173.7	167.5	161.5	155.5	149.7	144.1	138.8	133.6	128.6	123.8	119.2	114.8	110.4	106.2
54	180	173.8	167.7	161.7	155.8	150.0	144.5	139.1	134.0	129.0	124.2	119.5	115.0	110.7	106.4
55	180	173.9	167.9	161.9	156.0	150.3	144.8	139.5	134.3	129.3	124.5	119.8	115.3	110.9	106.6
56	180	174.0	168.0	162.1	156.3	150.6	145.1	139.8	134.7	129.7	124.8	120.1	115.6	111.1	106.8
57	180	174.0	168.1	162.3	156.5	150.9	145.5	140.1	135.0	130.0	125.1	120.4	115.8	111.4	107.0
58	180	174.1	168.3	162.5	156.8	151.2	145.8	140.5	135.3	130.3	125.4	120.7	116.1	111.6	107.2

TABLE A8.3 ELEVATION (ϕ_d POSITIVE)

Latitude, θ_T, degrees	LONGITUDE DIFFERENCE, ϕ_d, degrees														
	0	5	10	15	20	25	30	35	40	45	50	55	60	65	70
38	46.0	45.7	44.7	43.2	41.2	38.8	36.0	32.9	29.6	26.1	22.4	18.6	14.8	10.9	7.0
39	44.8	44.5	43.7	42.2	40.3	37.9	35.2	32.2	28.9	25.5	21.9	18.2	14.4	10.6	6.8
40	43.7	43.4	42.6	41.2	39.3	37.0	34.4	31.5	28.3	24.9	21.4	17.8	14.1	10.3	6.5
41	42.6	42.3	41.5	40.2	38.4	36.1	33.6	30.7	27.6	24.3	20.9	17.3	13.7	10.0	6.3
42	41.5	41.2	40.4	39.1	37.4	35.2	32.7	30.0	26.9	23.7	20.4	16.9	13.3	9.7	6.1
43	40.4	40.1	39.4	38.1	36.4	34.3	31.9	29.2	26.3	23.1	19.9	16.5	13.0	9.4	5.8
44	39.3	39.0	38.3	37.1	35.4	33.4	31.1	28.5	25.6	22.5	19.3	16.0	12.6	9.1	5.6
45	38.2	37.9	37.2	36.1	34.5	32.5	30.3	27.7	24.9	21.9	18.8	15.5	12.2	8.8	5.3
46	37.1	36.8	36.1	35.0	33.5	31.6	29.4	26.9	24.2	21.3	18.3	15.1	11.8	8.5	5.1
47	36.0	35.7	35.1	34.0	32.5	30.7	28.6	26.2	23.5	20.7	17.7	14.6	11.4	8.1	4.8
48	34.9	34.7	34.0	33.0	31.6	29.8	27.7	25.4	22.8	20.1	17.2	14.1	11.0	7.8	4.6
49	33.8	33.6	33.0	31.9	30.6	28.9	26.9	24.6	22.1	19.4	16.6	13.7	10.6	7.5	4.3
50	32.7	32.5	31.9	30.9	29.6	28.0	26.0	23.8	21.4	18.8	16.0	13.2	10.2	7.1	4.0
51	31.6	31.4	30.8	29.9	28.6	27.0	25.2	23.0	20.7	18.2	15.5	12.7	9.8	6.8	3.7
52	30.5	30.3	29.8	28.9	27.6	26.1	24.3	22.2	20.0	17.5	14.9	12.2	9.3	6.4	3.5
53	29.4	29.3	28.7	27.9	26.7	25.2	23.4	21.4	19.2	16.9	14.3	11.7	8.9	6.1	3.2
54	28.3	28.2	27.7	26.8	25.7	24.3	22.6	20.6	18.5	16.2	13.7	11.2	8.5	5.7	2.9
55	27.3	27.1	26.6	25.8	24.7	23.3	21.7	19.8	17.8	15.5	13.2	10.7	8.1	5.4	2.6
56	26.2	26.0	25.6	24.8	23.7	22.4	20.8	19.0	17.0	14.9	12.6	10.1	7.6	5.0	2.3
57	25.1	25.0	24.5	23.8	22.8	21.5	20.0	18.2	16.3	14.2	12.0	9.6	7.2	4.6	2.0
58	24.1	23.9	23.5	22.8	21.8	20.6	19.1	17.4	15.6	13.5	11.4	9.1	6.7	4.3	1.7

TABLE A8.4 ELEVATION (ϕ_d NEGATIVE)

Latitude, θ_r, degrees	LONGITUDE DIFFERENCE, ϕ_d, degrees														
	−0	−5	−10	−15	−20	−25	−30	−35	−40	−45	−50	−55	−60	−65	−70
38	46.0	45.7	44.7	43.2	41.2	38.8	36.0	32.9	29.6	26.1	22.4	18.6	14.8	10.9	7.0
39	44.8	44.5	43.7	42.2	40.3	37.9	35.2	32.2	28.9	25.5	21.9	18.2	14.4	10.6	6.8
40	43.7	43.4	42.6	41.2	39.3	37.0	34.4	31.5	28.3	24.9	21.4	17.8	14.1	10.3	6.5
41	42.6	42.3	41.5	40.2	38.4	36.1	33.6	30.7	27.6	24.3	20.9	17.3	13.7	10.0	6.3
42	41.5	41.2	40.4	39.1	37.4	35.2	32.7	30.0	26.9	23.7	20.4	16.9	13.3	9.7	6.1
43	40.4	40.1	39.4	38.1	36.4	34.3	31.9	29.2	26.3	23.1	19.9	16.5	13.0	9.4	5.8
44	39.3	39.0	38.3	37.1	35.4	33.4	31.1	28.5	25.6	22.5	19.3	16.0	12.6	9.1	5.6
45	38.2	37.9	37.2	36.1	34.5	32.5	30.3	27.7	24.9	21.9	18.8	15.5	12.2	8.8	5.3
46	37.1	36.8	36.1	35.0	33.5	31.6	29.4	26.9	24.2	21.3	18.3	15.1	11.8	8.5	5.1
47	36.0	35.7	35.1	34.0	32.5	30.7	28.6	26.2	23.5	20.7	17.7	14.6	11.4	8.1	4.8
48	34.9	34.7	34.0	33.0	31.6	29.8	27.7	25.4	22.8	20.1	17.2	14.1	11.0	7.8	4.6
49	33.8	33.6	33.0	31.9	30.6	28.9	26.9	24.6	22.1	19.4	16.6	13.7	10.6	7.5	4.3
50	32.7	32.5	31.9	30.9	29.6	28.0	26.0	23.8	21.4	18.8	16.0	13.2	10.2	7.1	4.0
51	31.6	31.4	30.8	29.9	28.6	27.0	25.2	23.0	20.7	18.2	15.5	12.7	9.8	6.8	3.7
52	30.5	30.3	29.8	28.9	27.6	26.1	24.3	22.2	20.0	17.5	14.9	12.2	9.3	6.4	3.5
53	29.4	29.3	28.7	27.9	26.7	25.2	23.4	21.4	19.2	16.9	14.3	11.7	8.9	6.1	3.2
54	28.3	28.2	27.7	26.8	25.7	24.3	22.6	20.6	18.5	16.2	13.7	11.2	8.5	5.7	2.9
55	27.3	27.1	26.6	25.8	24.7	23.3	21.7	19.8	17.8	15.5	13.2	10.7	8.1	5.4	2.6
56	26.2	26.0	25.6	24.8	23.7	22.4	20.8	19.0	17.0	14.9	12.6	10.1	7.6	5.0	2.3
57	25.1	25.0	24.5	23.8	22.8	21.5	20.0	18.2	16.3	14.2	12.0	9.6	7.2	4.6	2.0
58	24.1	23.9	23.5	22.8	21.8	20.6	19.1	17.4	15.6	13.5	11.4	9.1	6.7	4.3	1.7

TABLE A8.5 POLARIZATION OFFSET (degrees)

Latitude, θ_r, degrees	LONGITUDE DIFFERENCE, ϕ_d, degrees														
	0	5	10	15	20	25	30	35	40	45	50	55	60	65	70
38	0	6	13	18	24	28	33	36	39	42	44	46	48	49	50
39	0	6	12	18	23	28	32	35	38	41	43	45	47	48	49
40	0	6	12	17	22	27	31	34	38	40	42	44	46	47	48
41	0	6	11	17	22	26	30	33	37	39	41	43	45	46	47
42	0	6	11	16	21	25	29	33	36	38	40	42	44	45	46
43	0	5	11	16	20	24	28	32	35	37	39	41	43	44	45
44	0	5	10	15	20	24	27	31	34	36	38	40	42	43	44
45	0	5	10	15	19	23	27	30	33	35	38	39	41	42	43
46	0	5	10	14	18	22	26	29	32	34	37	38	40	41	42
47	0	5	9	14	18	22	25	28	31	33	36	37	39	40	41
48	0	5	9	13	17	21	24	27	30	33	35	36	38	39	40
49	0	4	9	13	17	20	24	27	29	32	34	36	37	38	39
50	0	4	8	12	16	20	23	26	28	31	33	35	36	37	38
51	0	4	8	12	16	19	22	25	28	30	32	34	35	36	37
52	0	4	8	11	15	18	21	24	27	29	31	33	34	35	36
53	0	4	8	11	15	18	21	23	26	28	30	32	33	34	35
54	0	4	7	11	14	17	20	23	25	27	29	31	32	33	34
55	0	4	7	10	14	17	19	22	24	26	28	30	31	32	33
56	0	3	7	10	13	16	19	21	23	26	27	29	30	31	32
57	0	3	6	10	13	15	18	20	23	25	26	28	29	31	31
58	0	3	6	9	12	15	17	20	22	24	26	27	28	30	30

A8.3 Polarization Offset (Sect.8.1.4)
The formula for calculation of polarization angle and therefore the polarization offset required to counteract the effect is more than a little complicated. However, with insignificant loss of accuracy, it can be reduced to a simple one:

$$\text{polarization offset} = \tan^{-1}\left(\frac{\sin \phi_d}{\tan \theta_r}\right) \text{degrees} \quad (8)$$

where ϕ_d is the longitudinal difference between satellite and receiving locations (Sect.8.1.2) and θ_r is the latitude of the receiving location.

A range of polarization offset values for Europe is given in Table A8.5 and its use is discussed in Section 8.1.4.

A8.4 Antenna Gain (Sect.8.2.2)
From Figure 8.9 in the main text, the radiation from an isotropic antenna is indicated by:

power flux density at a distance, r metres

$$\text{pfd} = \frac{P_t}{4\pi r^2} \quad (9)$$

where P_t is the power supplied. The figure of merit for a practical antenna is expressed by the number of times its radiated power in a given direction is greater than that of an isotropic antenna. It is denoted by G and is called the *gain* of the antenna. Hence:

$$G = \frac{\text{power radiated by antenna}}{\text{power radiated by istropic antenna}} \quad (10)$$

For a transmitting antenna with a gain G_t therefore, the radiated power flux density, pfd, at a distance r metres:

$$\text{pfd} = G_t \frac{P_t}{4\pi r^2} \text{ W/m}^2 \quad (11)$$

A receiving antenna excited by this power flux density will produce at its terminals a power P_r such that:

$$P_r = \text{pfd} \times A_{eff} \quad (12)$$

where A_{eff} is known as the *effective area* of the antenna. $A_{eff} = \eta A$ where A is the physical area and η the antenna efficiency. A_{eff} is related to the gain by the basic antenna formula:

$$A_{eff} = G \cdot \frac{\lambda^2}{4\pi}$$

Hence:

$$G = \frac{4\pi A_{eff}}{\lambda^2} \quad (13)$$

Also

$$A_{eff} = 10 \log(\eta A) \text{ dB} \quad (14)$$

where η is expressed as a fraction. This shows that G varies directly with A.

For a circular dish

$$A = \frac{\pi D^2}{4}$$

where D is the diameter and working in frequency in preference to wavelength,

$$G = \frac{\pi^2 D^2 f^2 \eta}{c^2} \quad (15)$$

where c = wave velocity, which, by entering D in cm, η in percentage, f in GHz, can be reduced to:

$$G \approx 0.00011\, D^2 f^2 \eta \quad (16)$$

A BASIC program for calculation of G and also in decibels follows (see program notes of Sect.8.1.2). Here is a typical print-out:

```
run
diameter of dish in centimetres . . . 120
efficiency of dish as a percentage . . . 55
frequency of transmission in GHz . . . 11.86

Gain of antenna = 12216 = 40.9 dB

Wavelength of transmission = 2.53 centimetres.
```

```
10 REM DISHGAIN
20 REM CALCULATION OF GAIN OF PARABOLIC DISH ANTENNA
30 INPUT "diameter of dish in centimetres......",d
40 INPUT "efficiency of dish as a percentage...",eta
50 INPUT "frequency of transmission in GHz.....",f
60 pi=3.14159: c=3*10^10
70 lambda=ROUND(c/(f*10^9),2)
80 G=(pi*pi*d*d*f*f*(10^18)*(eta/100))/(c*c)
90 gain=INT(G)
100 x=ROUND(10*LOG10(G),1): PRINT : PRINT
110 PRINT "Gain of antenna    = ";gain   " = ";x "dB": PRINT
120 PRINT "Wavelength of transmission  = ";lambda "centimetres"
```

(Continued on page 88)

TABLE A8.6 HYPOTHETICAL GAIN OF PARABOLIC DISH ANTENNA (dB)

FREQ. (GHz)	WAVE-LENGTH (cm)	DISH DIAMETER (cm)										
		30	60	70	80	90	100	110	120	130	140	150
EFFICIENCY = 50%												
11.7	2.56	28.3	34.3	35.7	36.8	37.8	38.8	39.6	40.3	41.0	41.7	42.3
11.8	2.54	28.4	34.4	35.7	36.9	37.9	38.8	39.7	40.4	41.1	41.8	42.3
11.9	2.52	28.5	34.5	35.8	37.0	38.0	38.9	39.7	40.5	41.2	41.8	42.4
12.0	2.50	28.5	34.5	35.9	37.0	38.1	39.0	39.8	40.6	41.3	41.9	42.5
12.1	2.48	28.6	34.6	35.9	37.1	38.1	39.0	39.9	40.6	41.3	42.0	42.6
12.2	2.46	28.7	34.7	36.0	37.2	38.2	39.1	39.9	40.7	41.4	42.0	42.6
12.3	2.44	28.7	34.8	36.1	37.3	38.3	39.2	40.0	40.8	41.5	42.1	42.7
12.4	2.42	28.8	34.8	36.2	37.3	38.3	39.3	40.1	40.8	41.5	42.2	42.8
12.5	2.40	28.9	34.9	36.2	37.4	38.4	39.3	40.2	40.9	41.6	42.3	42.9
EFFICIENCY = 55%												
11.7	2.56	28.7	34.7	36.1	37.2	38.3	39.2	40.0	40.8	41.4	42.1	42.7
11.8	2.54	28.8	34.8	36.1	37.3	38.3	39.2	40.1	40.8	41.5	42.2	42.8
11.9	2.52	28.9	34.9	36.2	37.4	38.4	39.3	40.1	40.9	41.6	42.2	42.8
12.0	2.50	28.9	35.0	36.3	37.4	38.5	39.4	40.2	41.0	41.7	42.3	42.9
12.1	2.48	29.0	35.0	36.4	37.5	38.5	39.5	40.3	41.0	41.7	42.4	43.0
12.2	2.46	29.1	35.1	36.4	37.6	38.6	39.5	40.4	41.1	41.8	42.5	43.1
12.3	2.44	29.2	35.2	36.5	37.7	38.7	39.6	40.4	41.2	41.9	42.5	43.1
12.4	2.42	29.2	35.2	36.6	37.7	38.8	39.7	40.5	41.3	42.0	42.6	43.2
12.5	2.40	29.3	35.3	36.6	37.8	38.8	39.7	40.6	41.3	42.0	42.7	43.3
EFFICIENCY = 60%												
11.7	2.56	29.1	35.1	36.4	37.6	38.6	39.5	40.4	41.1	41.8	42.5	43.1
11.8	2.54	29.2	35.2	36.5	37.7	38.7	39.6	40.4	41.2	41.9	42.5	43.1
11.9	2.52	29.2	35.3	36.6	37.8	38.8	39.7	40.5	41.3	42.0	42.6	43.2
12.0	2.50	29.3	35.3	36.7	37.8	38.9	39.8	40.6	41.3	42.0	42.7	43.3
12.1	2.48	29.4	35.4	36.7	37.9	38.9	39.8	40.7	41.4	42.1	42.8	43.4
12.2	2.46	29.5	35.5	36.8	38.0	39.0	39.9	40.7	41.5	42.2	42.8	43.4
12.3	2.44	29.5	35.5	36.9	38.0	39.1	40.0	40.8	41.6	42.3	42.9	43.5
12.4	2.42	29.6	35.6	37.0	38.1	39.1	40.1	40.9	41.6	42.3	43.0	43.6
12.5	2.40	29.7	35.7	37.0	38.2	39.2	40.1	40.9	41.7	42.4	43.0	43.6

Continued

TABLE A8.6 CONTINUED

FREQ. (GHz)	WAVE LENGTH (cm)	DISH DIAMETER (cm)										
		30	60	70	80	90	100	110	120	130	140	150
EFFICIENCY = 65%												
11.7	2.56	29.4	35.5	36.8	38.0	39.0	39.9	40.7	41.5	42.2	42.8	43.4
11.8	2.54	29.5	35.5	36.9	38.0	39.1	40.0	40.8	41.6	42.2	42.9	43.5
11.9	2.52	29.6	35.6	36.9	38.1	39.1	40.0	40.9	41.6	42.3	43.0	43.6
12.0	2.50	29.7	35.7	37.0	38.2	39.2	40.1	40.9	41.7	42.4	43.0	43.6
12.1	2.48	29.7	35.7	37.1	38.2	39.3	40.2	41.0	41.8	42.5	43.1	43.7
12.2	2.46	29.8	35.8	37.2	38.3	39.3	40.3	41.1	41.8	42.5	43.2	43.8
12.3	2.44	29.9	35.9	37.2	38.4	39.4	40.3	41.2	41.9	42.6	43.3	43.8
12.4	2.42	30.0	36.0	37.3	38.5	39.5	40.4	41.2	42.0	42.7	43.3	43.9
12.5	2.40	30.0	36.0	37.4	38.5	39.6	40.5	41.3	42.1	42.7	43.4	44.0
EFFICIENCY = 70%												
11.7	2.56	29.8	35.8	37.1	38.3	39.3	40.2	41.1	41.8	42.5	43.2	43.8
11.8	2.54	29.8	35.9	37.2	38.4	39.4	40.3	41.1	41.9	42.6	43.2	43.8
11.9	2.52	29.9	35.9	37.3	38.4	39.5	40.4	41.2	42.0	42.7	43.3	43.9
12.0	2.50	30.0	36.0	37.4	38.5	39.5	40.4	41.3	42.0	42.7	43.4	44.0
12.1	2.48	30.1	36.1	37.4	38.6	39.6	40.5	41.3	42.1	42.8	43.4	44.0
12.2	2.46	30.1	36.2	37.5	38.7	39.7	40.6	41.4	42.2	42.9	43.5	44.1
12.3	2.44	30.2	36.2	37.6	38.7	39.7	40.7	41.5	42.2	42.9	43.6	44.2
12.4	2.42	30.3	36.3	37.6	38.8	39.8	40.7	41.6	42.3	43.0	43.7	44.3
12.5	2.40	30.3	36.4	37.7	38.9	39.9	40.8	41.6	42.4	43.1	43.7	44.3
EFFICIENCY = 75%												
11.7	2.56	30.1	36.1	37.4	38.6	39.6	40.5	41.4	42.1	42.8	43.5	44.1
11.8	2.54	30.1	36.2	37.5	38.7	39.7	40.6	41.4	42.2	42.9	43.5	44.1
11.9	2.52	30.2	36.2	37.6	38.7	39.8	40.7	41.5	42.3	43.0	43.6	44.2
12.0	2.50	30.3	36.3	37.7	38.8	39.8	40.7	41.6	42.3	43.0	43.7	44.3
12.1	2.48	30.4	36.4	37.7	38.9	39.9	40.8	41.6	42.4	43.1	43.7	44.3
12.2	2.46	30.4	36.5	37.8	39.0	40.0	40.9	41.7	42.5	43.2	43.8	44.4
12.3	2.44	30.5	36.5	37.9	39.0	40.0	41.0	41.8	42.5	43.2	43.9	44.5
12.4	2.42	30.6	36.6	37.9	39.1	40.1	41.0	41.9	42.6	43.3	44.0	44.6
12.5	2.40	30.6	36.7	38.0	39.2	40.2	41.1	41.9	42.7	43.4	44.0	44.6

Continued

TABLE A8.6 CONTINUED

FREQ. (GHz)	WAVE-LENGTH (cm)	DISH DIAMETER (cm)										
		30	60	70	80	90	100	110	120	130	140	150
EFFICIENCY = 80%												
11.7	2.56	30.4	36.4	37.7	38.9	39.9	40.8	41.6	42.4	43.1	43.7	44.3
11.8	2.54	30.4	36.4	37.8	38.9	40.0	40.9	41.7	42.5	43.2	43.8	44.4
11.9	2.52	30.5	36.5	37.9	39.0	40.0	41.0	41.8	42.5	43.2	43.9	44.5
12.0	2.50	30.6	36.6	37.9	39.1	40.1	41.0	41.9	42.6	43.3	44.0	44.6
12.1	2.48	30.6	36.7	38.0	39.2	40.2	41.1	41.9	42.7	43.4	44.0	44.6
12.2	2.46	30.7	36.7	38.1	39.2	40.3	41.2	42.0	42.8	43.5	44.1	44.7
12.3	2.44	30.8	36.8	38.1	39.3	40.3	41.2	42.1	42.8	43.5	44.2	44.8
12.4	2.42	30.9	36.9	38.2	39.4	40.4	41.3	42.1	42.9	43.6	44.2	44.8
12.5	2.40	30.9	36.9	38.3	39.4	40.5	41.4	42.2	43.0	43.7	44.3	44.9

From this program suitably modified, Table A8.6 has been produced for estimation of gain values.

The Table is also useful for assessing the gain or loss arising from a change in dish diameter, given the transmission frequency and dish efficiency. As an example, there is an 8 dB gain in changing from a 60 cm diameter dish to 150 cm. This is to be expected since from Equation A8(16), G varies directly as D^2, hence:

$$\frac{G_{150}}{G_{60}} = 10 \log \frac{150^2}{60^2} = 7.96 \text{ dB} .$$

A8.5 Satellite Distance (Sect.8.3.2)
When latitude and longitude differences (θ_r and ϕ_r – Sect.A8.2) are both zero, then the satellite is directly overhead and its distance away, d is equal to h, the vertical height. As θ_r and/or ϕ_d increase, d increases also and the formula corrects for this by multiplying h by a factor slightly greater than 1. The formula is:

$$d = h\sqrt{1 + 0.42(1 - \cos\theta_r \, . \, \cos\phi_d)}$$

$$\therefore \quad d = 35786\sqrt{1 + 0.42(1 - \cos\theta_r \, . \, \cos\phi_d)} \text{ km} \qquad (17)$$

The BASIC computer program of Section A8.2 contains this formula and prints out the value of d for a range of values of θ_r, ϕ_r and ϕ_s.

A8.6 Free Space Path Loss (Sect.8.3.2)
This is a function mainly of satellite distance, d and to a lesser extent of frequency f (or wavelength, λ). From Equation (9) in this Appendix:

$$\text{pfd} = \frac{P_t}{4\pi d^2}$$

and it can be shown that the effective absorbing area of an isotropic receiving antenna is equal to

$$\frac{\lambda^2}{4\pi} .$$

Hence the available power at the terminals of the antenna:

$$P_r = \frac{P_t}{4\pi d^2} \times \frac{\lambda^2}{4\pi}$$

The basic transmission loss is equal to

$$\frac{P_t}{P_r} \quad \text{i.e.} \quad \frac{(4\pi d)^2}{\lambda^2}$$

(d and λ in metres). In decibels this is:

$$L_{fs} = 20 \log 4\pi d - 20 \log \lambda \text{ dB} \qquad (18)$$

Expressed in terms of frequency in GHz and d in kilometres:

$$L_{fs} = 92.44 + 20(\log f + \log d) \text{ dB} \qquad (19)$$

A8.7 Power Flux Density on Earth (Sect.8.3.3)
From the general formula:

$$\text{pfd} = \frac{\text{eirp}}{4\pi d^2} \qquad (20)$$

we get by changing to decibel notation and including an allowance for atmospheric loss:

$$\text{pfd} = \text{eirp} - 71 - 20 \log d - L_{at} \quad \text{dBW/m}^2 \qquad (21)$$

where d is in km, eirp in dBW and L_{at} is the atmospheric loss (dB).

A8.8 Receiving Dish Output (Sect.8.4)
From Equation A8(12), $P_r = \text{pfd} \times A_{eff}$:

Now $A_{eff} = \eta A$

where A is the area of the dish (m^2) which equals

$$\frac{\pi D^2}{4}$$

where D is the diameter in metres and η is a fraction.

Expressing in decibels and including the receiving losses, L_r

$$P_r = \text{pfd (dBW/m}^2) + 10 \log \left(\frac{\pi D^2 \eta}{4}\right) \text{(dB)} \qquad (22)$$

$$- L_r \text{ (dB)} \ \ldots \ \text{dBW}$$

The centre term is used as the basis for Table A8.7.

TABLE A8.7
EFFECTIVE AREAS OF CIRCULAR DISHES

dish diameter (cm)	A_{eff} expressed in decibels						
	efficiency, %						
	50	55	60	65	70	75	80
30	−14.5	−14.1	−13.7	−13.4	−13.1	−12.8	−12.5
40	−12.0	−11.6	−11.2	−10.9	−10.6	−10.3	−10.0
50	−10.1	−9.7	−9.3	−8.9	−8.6	−8.3	−8.0
60	−8.5	−8.1	−7.7	−7.4	−7.0	−6.7	−6.5
70	−7.2	−6.7	−6.4	−6.0	−5.7	−5.4	−5.1
80	−6.0	−5.6	−5.2	−4.9	−4.5	−4.2	−4.0
90	−5.0	−4.6	−4.2	−3.8	−3.5	−3.2	−2.9
100	−4.1	−3.6	−3.3	−2.9	−2.6	−2.3	−2.0
120	−2.5	−2.1	−1.7	−1.3	−1.0	−0.7	−0.4
140	−1.1	−0.7	−0.3	0	0.3	0.6	0.9
160	0	0.4	0.8	1.2	1.5	1.8	2.1
180	1.0	1.5	1.8	2.2	2.5	2.8	3.1

A8.9 Noise (Sect.8.5.1)
Of the sources of noise contributing to the total noise at the output of a satellite channel, only that due to thermal agitation can be described by a formula with any pretence to accuracy.

From Boltzmann's basic relationship between energy and temperature, the *available* noise power, P_n generated by thermal agitation of electrons is given by:

$$P_n = kTB \text{ watts} \qquad (23)$$

where

- k is Boltzmann's constant (1.38×10^{-23} watts per Hz per degree K)
- T is the temperature of the noise source in Kelvins
- B is the bandwidth (usually taken as that between the upper and lower 3 dB points of an amplifier).

Note the word "available" in P_n. It is the maximum power which can be obtained on the basis of matched conditions between noise source and load.

Satellite communication demands low noise because of the weak signals involved. For such systems the concept of *noise temperature* is found to be most convenient. It can apply to a single item or to a whole system. The Equivalent System Noise Temperature, T_s is the temperature to which a resistance connected to the input of a noise-free receiver would have to be raised in order to produce the same output noise power as the original receiver.

T_s is therefore an overall measure of noise performance and in the case of a satellite receiving installation can account for normal device thermal noise, radio noise picked up by the dish and thermal noise of the dish itself. Consider an LNB connected to a receiving dish as in Figure 8.13. The Figure is expanded to show the main elements as in Figure A8.2. The noise contribution of each unit is shown. P_{no}, the total noise power output is their sum. Then

$$P_{no} = GkT_sB$$

where G is the total gain. By equating this with the total noise power as shown in Figure A8.2 and rearranging:

$$T_s = \left(T_a + T_r + \frac{T_m}{G_r} + \frac{T_i}{G_rG_m}\right) \qquad (24)$$

showing that proceeding along the chain, the later stages have less effect, a not unexpected conclusion because they are followed by lower gain.

There is also a simple relationship between noise factor (Sect.8.5.2) and effective noise temperature, T_n for any device:

$$T_n = (F - 1)T \qquad (25)$$

where T is the reference temperature (usually 290 K) and F is the noise factor (at 290 K).

As an example, consider an LNB with a noise figure of 3.5 dB (= 2.24), then

$$T_n = (2.24 - 1) \times 290 = 360\text{ K}.$$

A8.9.1 G/T Ratios

In the main text, to avoid complication, only the term "signal-to-noise" ratio is used and it is shown just how important this is (Sect.2.7). It is generally considered that here, "signal" refers to the video waveform, not to that of the rf carrier. For Figure A8.2 therefore we should consider the *carrier-to-noise ratio* (C/N), i.e. the waveform before demodulation. From Equations (9), (12) and (13), for a receiving installation:

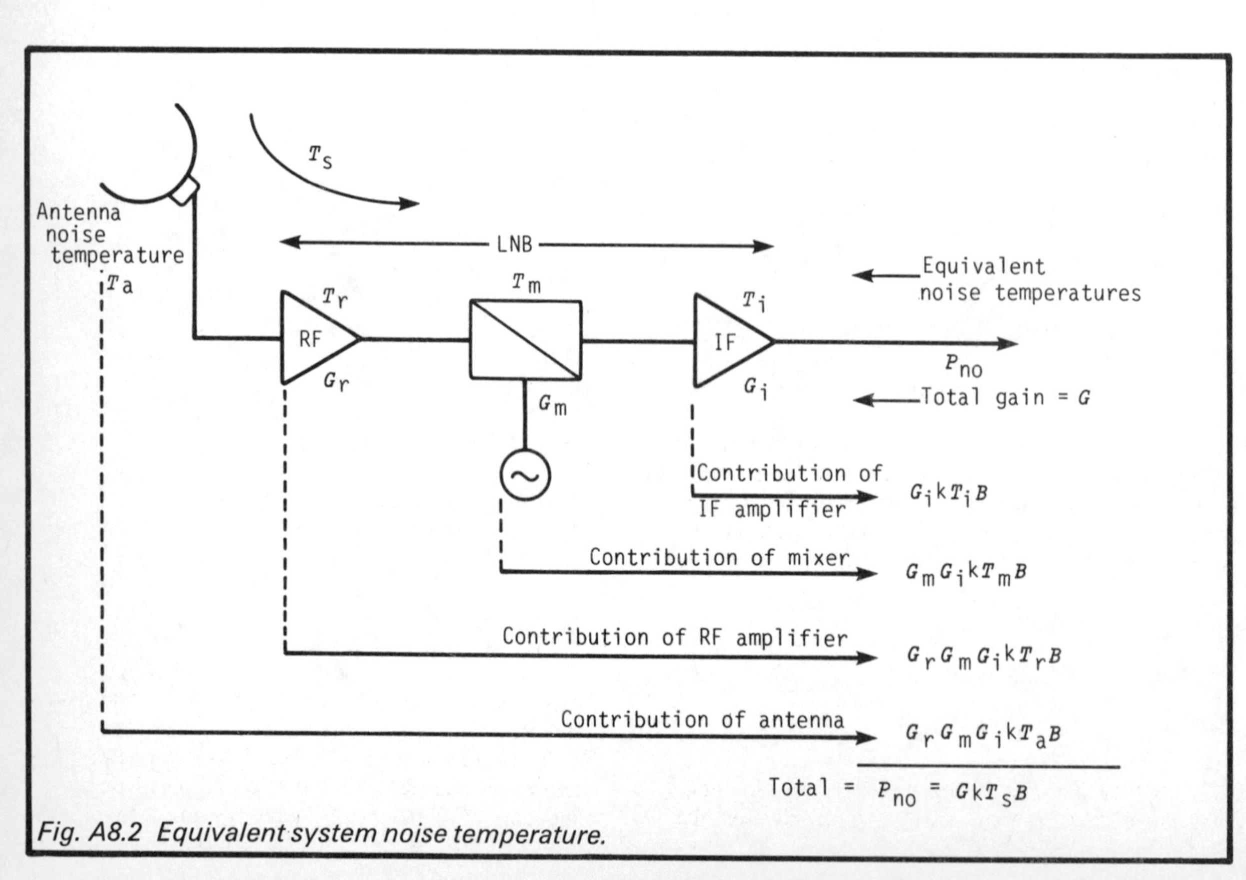

Fig. A8.2 Equivalent system noise temperature.

$$P_r = \frac{P_t G_t G_r}{4\pi d^2} \times \frac{\lambda^2}{4\pi}$$

$$\therefore \quad C/N = \frac{P_t G_t G_r}{4\pi d^2} \times \frac{\lambda^2}{4\pi} \times \frac{1}{kT_s B} \qquad (26)$$

All terms are constants except G_r and T_s hence $C/N \propto G_r/T_s$. The latter is generally known as the G/T ratio which is widely used as a figure of merit in assessing the overall performance of a receiving system as far as noise is concerned. With it the designer can study how gain can be traded for noise performance while still maintaining the same G/T ratio.

A8.9.2 Using C/N

As a very simplified example of the use of the C/N ratio, a noise power budget is developed below. This is only an illustrative example, many of the values can only be classed as typical and therefore do not apply to any particular situation.

Designers have access to data from which estimates of antenna equivalent noise temperature due to water, and oxygen in the atmosphere, sky noise etc. can be made (e.g. T_a at an elevation of 25° might be around 50 K). An LNB might have a noise temperature of some 300 K. Next suppose that for the whole receiving system, from Equation (24) the equivalent system noise temperature is calculated at, say, 400 K, then the noise power might be calculated as follows. Decibels again make life easier for which Boltzmann's constant becomes −228.6 dBW/K/Hz. From Equation (23):

k	−228.6 dBW/K/Hz
T_s (at 400 K)	26 dB (relative to 1 K)
B (at 27 MHz)	74.3 dB (relative to 1 Hz)
∴ P_n =	−128.3 dBW

(Note: B is generally taken as the IF bandwidth for this acts as a band-pass filter on the whole system.)

P_n is the total noise power, calculated via the system noise temperature. Looking back at the results of Table 8.1, for the DBS satellite,

$$C/N = \frac{\text{output signal power}}{P_n} = \frac{-109}{-128.3}$$

$$= 19.3 \text{ dB}$$

(but 16.3 dB if the receiving station is near the edge of the primary service area). Similarly for the EUTELSAT calculations:

$$C/N = \frac{-122}{-128.3} = 6.3 \text{ dB}$$

so it is evident that if a C/N of 14 dB is required (the range from good to approaching excellent is from about 12 to 17 dB), the DBS has plenty in hand for a rainy day, whereas, even with the larger dish, the particular EUTELSAT example has not. But it must be remembered that the calculations in Table 8.1 already take generous account of most losses.

A8.9.3 C/N to S/N

It is pertinent here to add a reminder that because the bandwidth of the satellite signal is around 27 MHz yet after demodulation the baseband is about 6–7 MHz, then system noise is reduced. This is simply deduced from the fact that as Equation (23) shows, the noise power is proportional to the bandwidth in which it is measured. The noise reduction does not follow a linear relationship however and the general formula linking C/N before demodulation of an fm wave to S/N at the output of the demodulator is

$$\frac{S/N}{C/N} = \frac{B}{f_m} \times 1.5 \left(\frac{\Delta f_{max}}{f_m}\right)^2 \qquad (27)$$

where B is the fm bandwidth, Δf_{max}, the maximum deviation and f_m, the maximum modulating frequency.

From Equations (2) and (3) of A4.1:

$$B = 2f_m(m + 1)$$

and since the modulation index,

$$m = \frac{\Delta f_{max}}{f_m},$$

then:

$$\frac{S/N}{C/N} = \frac{B}{f_m} \times 1.5\,m^2$$

$$\therefore \quad \frac{S/N}{C/N} = \frac{2f_m(m+1) \times 1.5\,m^2}{f_m} \qquad (28)$$

$$= 3m^2(m + 1)$$

hence, given only the value of m, the noise reduction can be calculated.

These formulae may be used to demonstrate the significant change in the ratio on demodulation. We can only take all-round figures for the PAL system but can rest assured that the full calculation gives a result several decibels better. Taking the bandwidth allowed (B) as 27 MHz and the maximum modulating frequency (f_m) restricted to 6 MHz, then from Appendix 4, Equation (3), $\Delta f_{max} = 7.5$ MHz and from Equation (2), $m = 1.25$. Then from Equation (28) above:

$$\frac{S/N}{C/N} = 10 \log 3 \times 1.25^2 \, (2.25) \approx 10 \text{ dB} .$$

Appendix 9

FORMULAE FOR CHAPTER 9

A9.1 Dish Declination Offset Angles (Sect.9.2.4)
The dish declination angle is illustrated in Figure 9.4 which shows that the sum of this angle and the elevation angle is equal to 90°. When a dish points due South then the difference in longitude, ϕ_d, between dish and pole of the arc is zero, hence from Section A8.2:

$$x = \cos^{-1}(\cos\theta \times \cos\phi_d)$$

but $\phi_d = 0 \quad \therefore \cos\phi_d = 1$

$$\therefore \quad x = \cos^{-1}(\cos\theta) = \theta$$

and from Equation (7), elevation is equal to

$$\tan^{-1}\left(\frac{1}{\tan\theta} - \frac{0.1513}{\sin\theta}\right)$$

$$= \tan^{-1}\left(\frac{\cos\theta - 0.1513}{\sin\theta}\right)$$

i.e. the elevation is related to a single variable, the latitude, θ.

If, using this formula, a graph is plotted of elevations over a range of latitudes for Europe (say, 36° – 60°) it is virtually a straight line conforming to the approximate relationship:

$$\text{Elevation} \approx 87.7 - 1.1 \times \text{latitude} \quad (1)$$

$$\therefore \quad \text{Dish declination} \approx 90° - \text{elevation}$$

$$\approx 2.3 + 1.1 \times \text{latitude} \quad (2)$$

Hence for any latitude a dish can be aligned by setting the elevation from Tables A8.3–8.4 or more easily by setting the declination from Equation (2).

To avoid calculation, what is most frequently used is a table of *dish declination offset angles.* These are angles which are added to the latitude to obtain the declination to which the dish is finally set. Table A9.1 gives the offset angles for European latitudes.

As an example, let us try out all the methods of calculating the setting-up angles for a polar-mount dish in, say, Brighton (UK) at approximately 51°N.

(i) From Table A8.3, elevation = 31.6°.

(ii) From Equation (1), elevation = 31.6°.

If the dish is calibrated for declination, then:

(iii) From Equation (2), declination = 58.4°.

(iv) From Table A9.1,

declination = 51 + 7.5 = 58.5°,

a reasonably consistent set of results.

TABLE A9.1
DISH DECLINATION OFFSET ANGLES

Latitude (degrees)	Offset angle (degrees)	Latitude (degrees)	Offset angle (degrees)
36	5.8	48	7.2
37	6.0	49	7.3
38	6.1	50	7.4
39	6.2	51	7.5
40	6.3	52	7.6
41	6.5	53	7.7
42	6.6	54	7.8
43	6.7	55	7.8
44	6.8	56	7.9
45	6.9	57	8.0
46	7.0	58	8.0
47	7.1	59	8.1
		60	8.2

Appendix 10

GLOSSARY OF SATELLITE TV TERMS

This is a list of terms and abbreviations readers may encounter in other satellite tv literature. To the experienced engineer some of the explanations may appear imprecise and even naive but this is deliberate in an attempt to help those less technically inclined.

AERIAL – see Antenna.

AFC – see Automatic Frequency Control.

AGC – see Automatic Gain Control.

ALTERNATING CURRENT (A.C.) – is an electric current which reverses its direction of flow at regular intervals. Radio waves produce alternating currents in antennas, speech and music produce alternating currents in a microphone.

AMPLITUDE – the strength or magnitude of a signal.

AMPLITUDE MODULATION – a method of impressing a signal on a carrier wave by varying the latter's amplitude. [Sect.4.4]

ANALOGUE – this usually refers to a mode of transmission of information. An analogue waveform has a physical similarity with the quantity it represents and therefore can usually be expressed by a graph on a base of time. Typical examples are given by the output of a microphone, tv camera or any device measuring a quantity which varies with time such as temperature, pressure etc.

ANTENNA – a device used to transmit or receive radio waves. Those used for satellite tv work at very high frequencies and are usually of the parabolic dish type. In the UK the term "aerial" has been and still is used.

APERTURE – as applied to an antenna is the area from which it radiates or receives energy.

APOGEE – the point farthest from Earth in the orbit of, for example, the Moon, a planet or an artificial satellite. A geostationary orbit is circular and therefore has no apogee. (See also "perigee".) [Sect.5.3]

APOGEE KICK MOTOR (AKM) – a rocket motor installed in a satellite which moves the satellite into its final orbit. [Sect.5.3]

ARIANE – the European expendable launch vehicle which puts satellites into orbit from the base at Kourou, French Guiana.

ASTRA – a 16-channel, medium power satellite operated by Société Européenne des Satellites (Luxembourg).

ATTENUATION – the reduction in amplitude of a signal through power losses in the channel over which it is travelling. [Sect.8.3.2]

ATTITUDE – position of spacecraft or satellite relative to specified directions. [Sect.6.1]

AUDIO – that which we can hear. The term is also used to describe the electrical representation of speech and music.

AUDIO FREQUENCY – any frequency of a sound wave which can normally be heard. The maximum range is from about 20 Hz to 20 kHz.

AUTOMATIC FREQUENCY CONTROL (AFC) – a circuit in a radio or tv receiver which ensures that the tuning circuits remain correctly adjusted to the incoming wave frequency.

AUTOMATIC GAIN CONTROL (AGC) – a circuit in a radio or tv receiver which maintains the output of an amplifying stage relatively constant irrespective of variations of the signal applied to the input (also known as "automatic volume control").

AZ/EL Mount – (Azimuth/Elevation). The basic parabolic antenna mount. Both azimuth and elevation are adjusted separately.

AZIMUTH – the horizontal angle measured from true North to the line joining an observer to a satellite. [Sect.8.1.2]

BAND-PASS FILTER – an electronic circuit which passes a pre-determined band of frequencies only. It does so by presenting high attenuation to all frequencies above and below the band, thereby preventing them from reaching the output terminals [Figs 4.3 and 9.5].

BANDWIDTH – the range between the highest and lowest frequencies in a communication channel, measured in hertz. [Sect.4.3]

BASEBAND – the range of frequencies initially generated and which is subsequently transmitted by radio (e.g. from audio or tv studios). [Sect.4.4]

BASIC – (Beginner's All-purpose Symbolic Instruction Code). A popular computer language, i.e. the series of special English-like instructions used in a program to tell the computer what to do.

BEAMWIDTH – a measure of the circle of sky which a receiving antenna "sees". Small beamwidths are better because the larger the circle, the greater the amount of sky noise picked up (see inset drawing on Fig.A6.1).

BINARY – of two. A numbering system which has two symbols only, generally designated by 0 and 1 [cf denary (decimal) which has 10 symbols (0, 1, . . . , 8, 9)].

BINARY CODE – a statement in binary digits. In computers a binary code is used to represent letters and instructions. Every code has an equivalent binary number. As an example, the letter W may be represented by the binary code 01010111 which also happens to be the binary number equivalent to 87 in decimal.

BIRD – a colloquial name for an artificial satellite.

BORESIGHT – the centre of a transmitting antenna beam.

BRIGHTNESS – see "luminance".

BROADCASTING SERVICE – a radiocommunication service for either sound or television for direct reception by the general public.

BUTTONHOOK FEED – it is possible to mount an LNB on a single rod emanating from the centre of a dish. Because the feedhorn looks back towards the dish, the guide has a buttonhook shape to ensure that the LNB is situated at the focal point.

CARRIER – a single-frequency radio wave which has impressed upon it a band of modulating frequencies (the baseband). [Sect.4.4]

CASSEGRAIN ANTENNA – an advanced form of parabolic antenna. It employs two reflecting surfaces as shown in Figure A10.1(i). This type is used mainly for ground transmitting and receiving stations (see also Gregorian antenna).

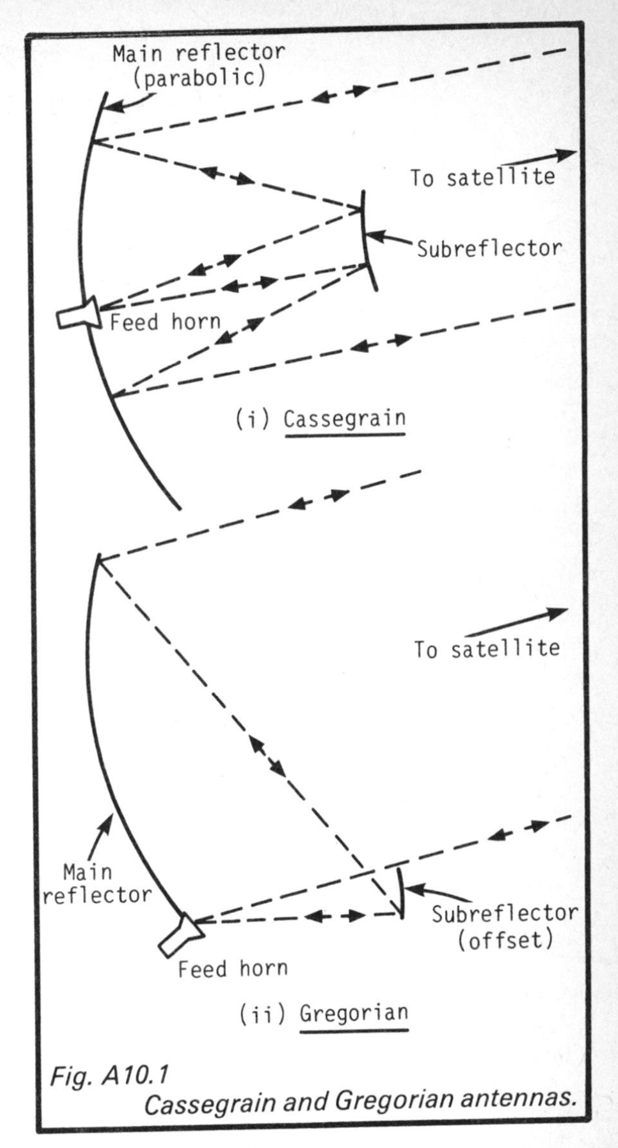

Fig. A10.1
Cassegrain and Gregorian antennas.

CATHODE-RAY TUBE (CRT) – an evacuated glass vessel in which an electron beam produces a luminous image on a fluorescent screen. [Sect.7.1]

C.A.T.V. – Community Antenna Television. Signals are received at a cable terminal and fed to subscribers over a cable network (Fig.1.2).

C-BAND – the satellite frequency band 4 – 6 GHz.

C.C.I.R. – French. In English this becomes "International Radio Consultative Committee". It is the international body which sets technical standards for radio transmissions. Operates under the auspices of the International Telecommunications Union.

C.C.I.T.T. – French. In English this becomes "International Telegraph and Telephone Committee". It is the international body which sets technical standards for telegraph and telephone systems. Operates under the auspices of the International Telecommunications Union.

CEEFAX – the teletext service of the BBC.

CHANNEL – the path over which information (data, audio, tv) is transferred. A channel can be set up on one or more of the following links: an air-path, pair of electrical conductors, terrestrial or satellite radio path, optical fibre.

CHROMINANCE – the part of a tv waveform containing the colour information. [Sect.7.4.1]

CIRCULAR POLARIZATION – a radio wave in which the electric and magnetic fields rotate as the wave travels. Facing an incoming wave, if the rotation is clockwise, the polarization is described as "Right Hand Circular" (RHC), if anti-clockwise it is "Left Hand Circular (LHC). [Sect.3.1.1]

CLARKE BELT – the geostationary orbit, named after the writer Arthur C. Clarke.

C/N – Carrier-to-noise ratio. This compares a radio carrier level with the noise level accompanying it. For satellite working C/N should be at least 14 dB. [Sect.A8.9.2]

COVERAGE AREA – the area of a footprint in which a tv picture may be satisfactorily received although freedom from interference from other overlapping transmissions is not guaranteed.

CROSS-COLOUR – a tv picture defect which results in swirling coloured patterns. [Sect.7.6]

CROSS-LUMINANCE – a tv picture defect which results in brightness variations at colour changes. [Sect.7.6]

CRYSTAL – (correctly termed "piezo-electric crystal"). A tiny piece of (usually) quartz which is capable of controlling the frequency of an oscillator [Sect.4.2] very accurately (e.g. quartz crystal clocks and watches).

CURRENT (electric) – the passage through a material of electrons. It is measured in *amperes* and for one ampere, in only one second, 6.25×10^{18} electrons pass by. One-tenth of one ampere flows through a hand-torch bulb, thousands of amperes are needed for an electric train.

DECIBEL – one-tenth of one *bel.* It is a unit used for comparison of power levels in electrical communication. [Sect.2.5]

DECLINATION – the angle at which a receiving antenna is set relative to the horizontal, used generally for the alignment of polar mounts. [Sect.9.2.4]

DECODER – see "Descrambler".

DECODING – the recovery of the original signal from a coded form of it.

DEMODULATION – the process in which a signal (the baseband) is regained from a modulated carrier wave. [Sect.4.4]

DENARY – of ten. The numerical system we use in daily life, more commonly known as *decimal.*

DEPOLARIZATION – the twisting of the polarization of a radio wave as it travels through the atmosphere. [Sect.3.1.1]

DESCRAMBLER – an electronic device which restores an encrypted signal to normal. [Sect.1.3.1]

DIGITAL – a method of handling information by measuring the amplitude of a quantity and coding that amplitude in the binary system.

DIRECT BROADCASTING BY SATELLITE (DBS) – a radiocommunication service mainly for television in which the signal is directed upwards towards a geostationary satellite which then re-transmits it for direct reception by the general public.

DIRECT CURRENT – an electric current which flows continuously in one direction only, for example as provided by a torch or car battery.

DISH – a commonly used name for a parabolic antenna. [Sect.8.1]

DISH ILLUMINATION – the area of a dish as "seen" by the feedhorn. The boundary of the area should coincide with the perimeter of the dish. If the area is greater, i.e. the feedhorn "sees" around the dish, noise is picked up; if smaller then signal pick-up is reduced.

DOWN CONVERTER – a frequency changer which converts a carrier wave to a lower frequency. [Sect. 8.1.3]

DOWNLINK – the radio channel from a satellite to Earth.

EARTH STATION – a ground-based transmitting or receiving installation working to satellites.

EBU – the European Broadcasting Union, an organization comprising the European national broadcasting authorities.

ECLIPSE – in satellite working, when the satellite passes through the shadow of the Earth and therefore is in darkness. [Sect.6.3]

E.C.S. – European Communications Satellites, these are satellites controlled by EUTELSAT.

EIRP – Effective Isotropically Radiated Power. The basis of a technique by which transmitted signal strength can be rated. An isotropic antenna is an elementary theoretical one considered to radiate from a point source equally in all directions. A practical antenna is then rated by its gain in a particular direction compared with the isotropic. [Sect. 8.2.1]

ELECTROMAGNET – consists of a coil of insulated wire wound round a metallic core. The latter becomes magnetized when an electric current flows in the coil but loses its magnetism when the current ceases. The magnetic field is used mainly to operate switches, also to rotate a polarization membrane within a parabolic dish feedhorn.

ELECTROMAGNETIC WAVE – the technical name for a radio wave, so called because it consists of electric and magnetic fields moving in unison. [Sect.3.1]

ELECTRON – an elementary particle carrying a charge of negative electricity. Electrons are present in all atoms and when in motion in one direction constitute an electric current. [Sect.2.8]

ELECTRON BEAM – a narrow pencil-like stream of electrons, all moving at very high velocity in one direction. The beam usually terminates on the screen of a cathode-ray tube and produces a spot of light.

ELECTRON GUN – an electronic device which generates free electrons and concentrates them into a beam. Used mainly in cathode-ray tubes. [Sect. 7.1]

ELEVATION – the angle between the horizontal and a line joining an observer to a satellite. [Sect.8.1.2]

ELV – Expendable Launch Vehicle. A rocket launcher which is used up or destroyed in flight and therefore is not available for re-use. [Sect.5.2]

ENCODING – expressing a message in code, e.g. when characters are stored in binary code in a computer. Often used to describe a process in which the form of an electronic signal is changed.

ENCRYPTION – the mutilation of a signal by the programme provider to prevent unauthorized use. The signal can be restored for authorized use.

E.S.A. – European Space Agency, the European organization responsible for research and satellites.

EUTELSAT – European Telecommunications Satellite Organization (Paris), the body responsible for EUTELSAT satellites. There are 26 member countries.

FEEDHORN – the device which collects the signal focussed onto it by a parabolic antenna [Fig.8.7].

FIELD – a sphere of influence in space or around us. A field is undetectable by the normal human senses yet is capable of creating a force and causing action (Sect.2.4.3). The most commonly encountered field is that of gravity which pulls bodies towards the centre of the Earth. *FIELD* is also a term used in television, it is the completion of a single picture scan. In interlaced scanning there are two fields to one complete picture [Fig.7.2].

FLICKER – the rapid increase and decrease of brightness of a tv picture. [Sect.7.2]

FLYBACK – the rapid movement of the spot on a tv screen from the end of one line or frame to the beginning of the next. [Sect.7.2]

FOOTPRINT – the area on land "illuminated" by a satellite radio beam. [Sect.1.1]

FRAME – a single, complete television image.

FREQUENCY – the number of cycles per unit time of an electric or electromagnetic signal. The unit is the hertz (one cycle per second). [Sect.2.6]

FREQUENCY MODULATION – a method of impressing a signal on a carrier wave by varying the wave frequency. [Sect.4.4]

GAIN – a measure of the increase in signal strength when it passes through a system. It can be expressed as the ratio of the power output of the system relative to the power input or more usually as this ratio in decibels. [Sect.4.1]

GEOSTATIONARY – stationary with respect to Earth. In the geostationary orbit satellites complete one revolution in the same time as the Earth does. Each satellite therefore moves so that it always remains above the same point on the Earth's surface. [Fig.1.1]

GIGAHERTZ (GHz) – a unit of frequency equal to one thousand million hertz (10^9 Hz).

GREGORIAN ANTENNA – an advanced form of parabolic antenna. It employs two reflecting surfaces and is usually offset as shown in Fig.A10.1(ii). Because there is no blocking of the radio wave by the subreflector, high efficiencies are obtained (see also "Cassegrain Antenna").

HALF POWER BEAMWIDTH (HPBW) – the beamwidth angle of a transmitting antenna which produces a footprint contour on which the signal power is 3 dB lower than the maximum value. [Sect.A6.1 and Fig.A6.1]

HARMONIC – a component of a wave having a frequency an integral number of times that of the basic (fundamental) frequency, e.g. if the fundamental frequency is denoted by f Hz, then its harmonics are $2f$, $3f$, $4f$, etc.

HDTV – High Definition Television – newer systems being developed having a greater number of lines than the current 625.

HERTZ – the international standard unit of frequency equal to one cycle per second (after Henrich Hertz, a German physicist).

HORIZONTAL POLARIZATION – a radio wave in which the electric field is horizontal and the magnetic field vertical. [Sect.3.1]

HORN (electromagnetic) – a horn-shaped (i.e. of expanding cross-section) termination on a waveguide used in both transmitting and receiving paraboloid antennas. Generally the orifice is of rectangular or circular shape [Fig.6.4].

INCLINOMETER – an instrument used to measure angles of elevation. [Sect.9.2.1]

INTELSAT – the International Telecommunications Satellite Organization, a body controlling the international satellite system for telephony, data and tv.

INTERMEDIATE FREQUENCY (IF) – a frequency to which that of a modulated carrier wave is reduced for processing [Fig.9.5].

ITU – International Telecommunications Union – concerned with international standards for radio, telegraph and telephone. It is the body responsible for allocating frequencies for satellite working to the various regions of the World.

KELVIN – a degree of temperature equal to a Centigrade or Celsius degree. The Kelvin scale however starts at absolute zero so 0° Celsius is equivalent to 273° Kelvin (after Lord Kelvin, a British physicist).

KINETIC ENERGY – the energy possessed by a body due to its weight and motion.

KU BAND – the satellite frequency band 11 – 14 GHz (Europe).

L-BAND – the frequency range 1 – 2 GHz (USA).

LED – Light-Emitting Diode. Usually a very small electric lamp having no filament. The glow is normally but not necessarily red. These small sausage-shaped lamps are used mainly as indicators or as the basic units of an illuminated letter or number display.

LHCP – left-hand circular polarization of a radio wave. Looking towards an oncoming wave, the rotation is anti-clockwise. [Sect.3.1.1]

LNA – low noise amplifier.

LNB – low noise block converter – an electronic device at the output of a receiving antenna which amplifies the incoming radio wave and converts it to a lower, intermediate frequency for transmission over a cable to the satellite receiver. [Sect.8.1.3]

LNC – low noise converter – see LNB.

LOOK ANGLE – the angle of elevation of a satellite. It varies both with position of observer and of satellite.

LOSS – a measure of the extent to which the amplitude of a signal is decreased by its passage through a system. Usually expressed in decibels.

LUMINANCE – the luminous intensity or amount of white light emitted from a small area on a tv screen. With regard to a tv waveform, it is that part which contains the brightness information. [Sect. 7.4]

MAC – Multiplexed Analogue Components, a partly analogue, partly digital system of tv transmission developed especially for DBS. MAC produces pictures of enhanced quality compared with the PAL system. [Sect.7.6]

MAGNETIC NORTH – the Earth can be considered as a huge magnet having North and South magnetic poles. The line joining these poles is inclined slightly to the axis of rotation hence true North and magnetic North do not coincide. A compass points to magnetic North. [Sect.9.2.1]

MAGNETIC VARIATION – the angular difference at any place between true North and magnetic North.

MASS – the quantity of matter a body contains as measured by its acceleration when a given force is applied. [Sect.2.4.2]

MEMORY – an electronic store for information. For example, a satellite receiver can store in its memory data with regard to tv channels such as frequency, azimuth, elevation, polarization.

MERIDIAN (true meridian) – a circle on the Earth's surface of constant longitude passing through a given place and the North and South poles.

MICROWAVES – ultra-short waves of wavelength less than about 30 cm (frequency = 1 GHz) [Fig. 2.2].

MIXER – an electronic device which accepts two different frequencies at its input and produces a combination of these frequencies at the output. [Sect.4.4]

MODULATION – the process in which a signal (the baseband) is impressed upon a higher frequency carrier wave. [Sect.4.4]

MULTIPLEX – transmission of several separate elements over one channel [e.g. Sect.7.6].

N.A.S.A. – National Aeronautics and Space Administration. The USA organization responsible for the space exploration programme.

NEGATIVE – the name given to the electrical charge of the electron. [Sect.2.8.1]

NOISE – any unwanted electrical or audio signal which accompanies but has no relevance to the transmitted signal. [Sect.2.7]

NOISE TEMPERATURE – a method of assessing electrical noise. It is the temperature in degrees Kelvin to which the noise source would have to be raised to produce the same noise output as the system itself. [Sect.8.5.2]

N.T.S.C. – National Television Standards Committee. This also refers to the 525 line/60 field system used for tv broadcasting in the USA and some other countries.

OFFSET ANGLE – the angle which is added to the latitude of a receiving installation for setting the declination angle to line up a polar mount with the geostationary arc (Clarke Belt).

OFFSET ANTENNA – a special design of parabolic antenna in which the LNB is outside of the path of the incoming signal.

OPTICAL FIBRE – a very fine circular strand of glass, about human hair thickness. Such a fibre transmits electromagnetic waves at frequencies of light with very high bandwidths. Optical fibres are used for the transmission of telephony, data and television.

ORACLE – the teletext service of the UK Independent Broadcasting Authority.

ORBIT – the path a satellite follows around a larger body, usually the Earth [Fig.1.1].

OSCILLATOR – an electronic device for the production of alternating electric currents, i.e. waveforms having frequencies from one or two up to many thousands of millions of reversals per second.

PAL – Phase Alternation Line. The 625 line/50 field tv system used by many European countries (except France).

PARABOLA – a curve obtained by slicing a cone at a certain angle (see Fig.8.2). It conforms to the mathematical equation $y^2 = 4fx$ where f is the distance of the focus from the centre (focal length). [Sect.A8.1]

PARABOLIC ANTENNA – a dish type of antenna having a curved surface of parabolic shape. This has the capability of reflecting an incoming radio wave and focussing its energy onto a single point. [Sect. 8.1.1]

PAYLOAD – the productive or useful part of the load of a rocket.

PERIGEE – the point nearest to Earth in the orbit of, for example, the Moon, a planet or an artificial satellite. A geostationary orbit is circular and therefore has no perigee. (See also "apogee".) [Sect.5.3]

PERSISTENCE OF VISION – an image in the eye persists for a period of time (up to about 0.1 seconds). [Sect.7.2]

PETALIZED – a form of construction of dish antennas by using identical metal "petals" bolted together to form the dish instead of the more usual solid construction.

PICTURE ELEMENT – the smallest area of a tv picture which can be displayed.

POLAR CURVE – one which is related in a particular way to a given curve and to a fixed point called a pole. The Sun follows a polar curve each day. [Sect. 9.1.1]

POLARIZATION – the way in which the electric field of a radio wave is disposed relative to the direction of propagation. [Sect.3.1.1]

POLAR MOUNT – a special type of parabolic antenna mounting which allows the antenna to rotate and at the same time adjust its elevation so as to follow the geostationary arc. [Sect.9.2.4]

POLAROTOR – an electronic device which rotates an LNB for correct alignment with the polarization of the incoming radio wave. Alternatively a membrane built into the feedhorn may be moved into the required positions by an electromagnet. [Sect.9.1.2]

POSITIVE – the name given to an electrical charge opposite to that of the electron. [Sect.2.8.1]

PRE-AMPLIFIER – an amplifier which raises a low-level signal to a value suitable for driving a main amplifier.

PRIMARY FOCUS ANTENNA – a parabolic dish with the LNB situated at the focus [Fig.8.3(i)].

PROPELLANT – fuel used in a rocket engine to provide thrust. [Sect.5.2]

RADIAN – the angle at the centre of a circle subtended by an arc equal in length to the radius. It is equivalent to 57.3 degrees.

RADIATION – the outflow of energy in the form of a radio wave, generally from a radio antenna.

RANDOM NOISE – electrical noise generated by the continual movements of a large number of free electrons in a conductor [Sect.8.5.1]. (See also "Thermal Noise".)

RASTER – a pattern of scanning lines on a tv screen. [Sect.7.2]

REPEATER – an electronic device for automatic re-transmission or amplification of a signal.

RHCP – right-hand circular polarization of a radio wave. Looking towards an oncoming wave, the rotation is clockwise. [Sect.3.1.1]

SATCOM – a series of satellites providing services in the USA.

SATELLITE – a heavenly or artificial body revolving round another larger one.

S-BAND – the frequency range 2–4 GHz (USA).

SCANNING – the resolution in a prearranged pattern of a tv picture into its elements of light, shade and colour. [Sect.7.2]

SCRAMBLING – see Encryption.

SECAM – Système En Couleurs à Mémoire, the 625 line/50 field tv system used in France, Luxembourg and Monaco.

SERVICE AREA – the area within a footprint in which the signal is sufficiently strong for a satisfactory and interference-free tv picture to be received.

SES – Société Européenne des Satellites – the consortium based in Luxembourg which is responsible for the ASTRA satellites.

SHUTTLE – the re-usable space vehicle manned by astronauts which is capable of carrying satellites up into space for launching into orbit. The choice for NASA's space programme. [Sect.5.2]

SIGNAL – an intelligble sign conveying information. In satellite tv signals are in the form of electronic waveforms varying with time. There are radio, video and audio signals.

SIGNAL LEVEL METER – a measuring device, usually employing a needle moving over a scale, which can be tuned to a signal. The travel of the needle across the scale indicates the strength of the signal applied.

SIGNAL-TO-NOISE RATIO – a method of indicating the strength of a signal compared with that of the noise accompanying it, usually expressed in decibels. [Sect.2.7]

SMATV – Satellite to Master Antenna Television. Signals are received on a single dish and fed by cable to apartments, throughout a hotel, group of houses, etc. [Fig.1.2].

SPARKLIES – a colloquial term for white dots or flashes on a tv screen.

SPOT BEAM – a narrow satellite radio beam. The footprint is therefore smaller than for a normal beam. [Sect.6.4.2]

STS – Space Transport System, the American space "Shuttle", a reusable launch vehicle.

TELEMETRY – taking readings of a measuring instrument at a distance, usually via a radio link. [Sect.6.1]

TELETEXT – separate information transmitted with a tv picture signal and which can be displayed on the screen in place of the normal picture. In the UK the BBC calls it CEEFAX, the IBA uses the name ORACLE.

TERRESTRIAL – of the Earth (from Latin, *terra* = earth).

THERMAL NOISE – electrical noise which arises from the agitation of electrons in a conductor due to heat. [Sect.8.5.1]

TIME BASE – electronic equipment for generating a repetitive timing voltage.

TONNE – a metric measurement of weight equal to 1000 kilograms (roughly the same as the Imperial "ton").

TRANSFER ORBIT – an orbit in which a satellite is first placed prior to being moved into the final orbit. [Sect.5.3]

TRANSMITTER – the equipment used to transmit speech, data, tv by radio wave.

TRANSPONDER – an electronic system in a satellite which receives a signal from Earth, changes its frequency, then amplifies it for transmission back to Earth. [Sect.6.2.1]

TRIGGER – a short duration electrical pulse which starts some action. [Sect.7.3]

TTC – Telemetry, Tracking and Command – these are the facilities required by a ground control station for monitoring the spatial and electrical conditions of a satellite and for signalling back instructions.

TUNED CIRCUIT – an electronic circuit which resonates or "tunes" to one particular frequency only.

TVRO – Television Receive Only – a single domestic satellite tv installation.

TWT – Travelling Wave Tube – a special electronic amplifier in which very high frequency waves travel along a tube and as they do so are increased in amplitude. Used in satellites. [Sect.6.2.1]

UPLINK – the radio channel from Earth up to a satellite [Fig.1.2].

VERTICAL POLARIZATION – a radio wave in which the electric field is vertical and the magnetic field horizontal. [Sect.3.1]

VIDEO – that which we can see. A term used in the broadcasting of television pictures. A video waveform is the band of frequencies representing the output of a tv camera. [Sect.7.4]

VIDEOTEX – a system in which signals are sent over telephone lines for ultimate display on the screen of a tv receiver (also known as "Viewdata").

VIEWDATA – see "Videotex".

VISION SIGNAL – in tv, a carrier wave modulated by the video signal or waveform.

VOLT – a measure of the difference of electric potential in a circuit. Car batteries are 12 volt, the electricity mains, over 200 volts, lightning is many millions.

WARC – World Administrative Radio Conference. A conference was called in 1977 to allocate satellite channels to the various nations of the World. Operates under the auspices of the International Telecommunications Union. [Sect.3.4]

WATT – the unit of electric power. Only one watt or so is required to light a hand torch bulb but several thousand watts are required for an electric cooker.

WAVEFORM – the shape given by plotting the amplitude of a varying quantity against time on a graph.

WAVEGUIDE – a metal tube of rectangular or circular cross-section through which microwaves can be transmitted. [Sect.3.5]

WIRE BROADCASTING – the distribution of sound and/or tv programmes to subscribers over a network of cables.

X-BAND – the frequency range 8–12.5 GHz (USA).

Appendix 11

STATE OF THE ART (JANUARY 1989)

From the tv point of view there are now several satellites of interest which are transmitting to Western Europe. They are communication satellites using spare capacity for tv distribution and are comparatively low powered, hence requiring the larger types of dish. These are:

EUTELSAT-1 F1 (13°E)
11 channels consisting of a mixture of European languages including several in English.

EUTELSAT-1 F2 (10°E)
2 channels, Norwegian and English.

EUTELSAT-1 F4 (10° E)
1 channel, Spanish only.

INTELSAT V F2 (1°W)
3 channels, Norwegian and Swedish.

INTELSAT VA F11 (27.5°W)
11 channels, 7 of which are in English. Also French, Dutch, Spanish and Scandinavian.

INTELSAT VA F12 (60°E)
7 channels, mainly in German.

TELECOM-1 F1 (5°W)
2 channels, both French.

Planned but not yet in service:

TV-SAT (19°W)
Deutsche Bundespost, West Germany.

TDF-1 (19°W)
Telediffusion de France.

Both the above are high powered DBS satellites.

ASTRA (19.2°E)
Société Européenne des Satellites (SES), Luxembourg.

ASTRA is a medium powered satellite (see Table 8.1) with 16 channels, most of which are expected to be in English. Transmissions are expected to commence in February 1989. The downlink frequency range is 11.20 – 11.45 GHz (below DBS). Generally 60 cm dishes are considered suitable except for the north of Britain, south of France, Spain, Italy and Scandinavia where 75 – 125 cm dishes will be needed to compensate for the lower power flux density on the ground of 2 – 4 dB.

Four channels have already been allocated to English programmes under the auspices of *News International.* They are to be broadcast directly in PAL and initially are to be funded by advertising, hence no encryption. Accordingly all four channels are receivable at present on existing television sets. If however in the future any channels are changed to D-MAC specification an add-on MAC decoder will be required. A special low-cost package which includes a 60 cm motorized dish with polarotor and satellite receiver is expected to be available.

These channels are to be beamed up to ASTRA by British Telecom from North Woolwich, London, instead of from the main TTC centre at Betzdorf (near Luxembourg).

BSB1 (31°W)
British Satellite Broadcasting Ltd.

This is the British DBS satellite scheduled for launch in August 1989 by Hughes Aircraft Company Delta rocket. A second satellite is to be in orbit about one year later as a back-up. Three channels will be on offer, Nos. 4, 8 and 12 (Sect.3.4), all encrypted and with D-MAC coding (Sect.7.6). Each of the three transponders has an output of 110 watts obtained from 2 x 55 watt travelling-wave amplifiers in parallel (Sect.6.2.1). The programme up-link and TTC centre is located at Chilworth, near Southampton (Hampshire).

Special low-cost or de-luxe packages will be available consisting of a 35 – 60 cm dish and receiver to work directly into an existing television set. Two additional features in the receiver will be stereo sound and D-MAC decoding in readiness for MAC-type tv sets.

For information on programmes and programme times, see any of the monthly/bi-monthly magazines mentioned in Appendix 1 (A1.2).

INDEX